ROUTLEDGE LIBRARY EDITIONS:
LOGIC

Volume 4

THE LOGICAL STRUCTURE OF SCIENCE

THE LOGICAL STRUCTURE OF SCIENCE

A. CORNELIUS BENJAMIN

LONDON AND NEW YORK

First published in 1936 by Kegan Paul, Trench, Trubner & Co. Ltd.

This edition first published in 2020
by Routledge
2 Park Square, Milton Park, Abingdon, Oxon OX14 4RN

and by Routledge
52 Vanderbilt Avenue, New York, NY 10017

Routledge is an imprint of the Taylor & Francis Group, an informa business

© 1936 A. Cornelius Benjamin

All rights reserved. No part of this book may be reprinted or reproduced or utilised in any form or by any electronic, mechanical, or other means, now known or hereafter invented, including photocopying and recording, or in any information storage or retrieval system, without permission in writing from the publishers.

Trademark notice: Product or corporate names may be trademarks or registered trademarks, and are used only for identification and explanation without intent to infringe.

British Library Cataloguing in Publication Data
A catalogue record for this book is available from the British Library

ISBN: 978-0-367-41707-9 (Set)
ISBN: 978-0-367-81582-0 (Set) (ebk)
ISBN: 978-0-367-41797-0 (Volume 4) (hbk)
ISBN: 978-0-367-42607-1 (Volume 4) (pbk)
ISBN: 978-0-367-85393-8 (Volume 4) (ebk)

Publisher's Note
The publisher has gone to great lengths to ensure the quality of this reprint but points out that some imperfections in the original copies may be apparent.

Disclaimer
The publisher has made every effort to trace copyright holders and would welcome correspondence from those they have been unable to trace.

THE LOGICAL STRUCTURE OF SCIENCE

BY

A. CORNELIUS BENJAMIN

*Assistant Professor of Philosophy,
The University of Chicago*

PSYCHE MONOGRAPHS: No. 9

KEGAN PAUL, TRENCH, TRUBNER & Co. Ltd.
Broadway House, Carter Lane, London, E.C.
1936

Printed in Great Britain by
R. I. SEVERS, CAMBRIDGE

PREFACE

It has become a commonplace in philosophy that a problem which is well defined is half solved. It is also a recognized principle of philosophy that a position can be adequately defined only when an alternative or opposing position endeavors to achieve formulation. From these two propositions it follows by a not too rigorous logic that when an issue has become sharply defined through the more or less precise formulations of the contrasting points of view, the way is open either for the complete overthrow of one of the positions or for the reconciliation of the opposition through a " higher synthesis."

Probably few would disagree with the assertion that the outstanding issue in the history of the philosophy of science is that between the positivists and the anti-positivists. But the issue, in spite of its venerable history, has not been precisely defined until comparatively recent times. The formulations of the positivists (Comte, Mill, Spencer, Avenarius, Mach, Poincaré, Pearson, Hobson) were essentially obscure, and the statements of their opponents (Haeckel, Büchner, Bergson, Eddington, Jeans) were no better. Recently, however, positivism has received new blood from the so-called Vienna group (Carnap, Schlick, Wittgenstein) and from Bridgman and the pragmatists in this country. At the same time anti-positivism has received stimulation from a somewhat more miscellaneous group containing Meyerson, Bavink, Whitehead, and Planck. In the hands of these more recent representatives of both positions the issue is being reformulated through the technique of logic and the general theory of sybolism. But as if this were not enough, there has arisen still another group (Russell, Broad, G. E. Moore, Stace, and many of the American new, and critical realists) who do not attack positivism and thus seem to ally themselves with this position but who call themselves realists upon the grounds of " construction " and thus seem to employ the method which is essentially that of positivism; just how to tag these realists who construct their world from immediately perceived data is somewhat of a problem.

There are many who will say that through these many formulations the issue has become essentially confused. This may be the case. But one conclusion, at least, seems to be justified, viz. that through these various attempts at defining the issue the *rapproachment* of the opposing views is being achieved. Hence the problem is being confused only in the sense that because of the strict continuity of the positions the unambiguous location of an exponent of either positon in one class rather than the other is impossible.

Most of all I note this difficulty with reference to myself. Frankly, I do not know whether to call myself a positivist or not. I shall take the liberty of calling the position which I am going to defend in the following pages " critical positivism." Labels are always dangerous and they are particularly so in this case. I ask the reader, therefore, to avoid, if possible, giving any content to this term until he has examined my formulation of the position. At this point I wish only to state my firm conviction that all science and philosophy must start with the given. But the range of the given is not definite. There are things which are clearly given and obvious, but there are also things which are obscurely given and conjectural. It is the task of science and philosophy to symbolize this *total* realm of the given, through the discovery of relations connecting the clearly given with the obscurely given. In terms of such relations we may *derive* the obscurely given from the clearly given and *explain* the clearly given by the obscurely given. The position which I shall defend is positivistic in its insistence that one should begin with the clearly given and that only in terms of this realm should one talk about the obscurely given. Thus it insists upon the essentiality of methods of construction and reduction. But it is anti-positivistic in its insistence that one should not give to all constructions and derivations an unreal or a non-existential status. Hence it does not deny hypothetical entities but merely gives them the status of obscure entities. Positivism fails to explain why we are not satisfied with mere description. I have tried to meet this difficulty by insisting that the obscure is part of the given. Crudely stated, the obscure is hinted at in the obvious.

The proper examination of the problem demands certain preliminary considerations. Hence I begin with an examination of science *from the outside*, i.e. as a type of human activity having a place in a general structure of activities of a similar kind. This is essentially the problem of the classification of the sciences. But I then turn to an examination of science *from the inside*, i.e. as a type of cognition exhibiting certain features. This will reveal the fundamental dependence of knowledge upon that which we are striving to know, and will necessitate a short excursion into metaphysics. Chapters III and IV are the results of this analysis. If the reader dislikes metaphysics he is advised to skim over these chapters, though a proper understanding of my position cannot be gained without taking them into consideration—especially the latter part of Chapter IV. The analysis of awareness in the following chapter continues the discussion of the second chapter. The attempt to formulate the character of operational acts in Chapter VI, I consider very important, since it is an attempt to make explicit a notion which has been implicit in much of recent philosophy of science. Chapters VII, VIII, and IX belong together since, as a whole, they constitute my general theory of symbolism. They attempt to show, respectively, how meanings in general are derived from occurrents, how correlational symbols are derived, and how suppositional symbols are derived. The last five chapters constitute a unit, the program of which is announced in Chapter X. The attempt will be made to show how one form of symbolism tends to supersede another as we strive to include more of the obscurely given. Models and pictorial symbols reveal a high degree of adequacy on the lowest level. Descriptive science illustrates symbolism on a secondary level. Explanatory science, with one of its most significant kinds, viz. quantitative explanation, illustrates the third level. The attempt will be made to show that the development from one stage to the next is inevitable, consequently that explanatory science represents the culmination of knowledge.

No one is more conscious than I of the limitations and inadequacies of my attempted solution of this important problem. The reader should not be misled by the dogmatism with which I have decided many highly controversial issues. One may always be more dictatorial in exposition than in defence. It is needless to say, therefore, that many of the solutions here offered are still highly conjectural in my own mind. My justification for presenting them before I have satisfied myself either as to what they mean or as to whether they are true is twofold—a discovery and a hope. The discovery is that through the process of formulating and organizing my ideas around a central thesis I find that they have taken on precision. The hope is that through the act of putting them before the public I may receive the benefits of the criticisms of others who are interested in the same problem. Thus I find that I know better what I mean than I should if I had not tried to express my ideas, and I hope that I shall know better whether they are true. If the ideas are premature, I offer this as my only apology.

The plan which I have adopted is somewhat unusual. In general I have made little or no reference in the body of the text to other writers upon the same problem, either to those with whom I agree or to those with whom I disagree. Such reference as there is occurs in footnotes, though even the latter are not nearly so extensive as the literature upon the subject warrants. My reason for this is that in order either to contrast or to compare my position with that of another I must first interpret the position of the other. In this interpretation I run the risk of being unfair to him. Apparently few philosophers are disturbed by this fact. But there is a more serious risk. In contrasting and comparing my position with my interpretation of the position of another I endanger the univocal character of my own exposition. For words take on meanings through all situations in which they are used. I have chosen, therefore, to let my words speak for themselves.

This results in a somewhat cumbersome style. Repetition of both words and structural forms is inevitable if one is concerned with precision. Avoidance of metaphors, analogies, and other figures of speech is also a necessary concomitant of accuracy. Ideally the exposition should have taken the form of Wittgenstein's *Tractatus Logico-Philosophicus*—but I could not quite persuade myself that such a forbidding form was necessary.

Special acknowledgments are due to Miss Albertine Loomis for correcting the entire manuscript and suggesting valuable improvements in style; to Dr. Lee Byrne for the preparation of the excellent index; and to my wife, not only for help in the correction of the manuscript, but for constant encouragement during the long period of its preparation. Grateful recognition is also due to the John Simon Guggenheim Memorial Foundation for financial assistance which permitted me to carry on studies of vital importance to the development of my general thesis; without this aid the manuscript could hardly have taken form.

Chicago, March 23rd, 1936.

TABLE OF CONTENTS

CHAPTER	PAGE
I. THE PLACE OF SCIENCE	15
Human Activities	17
Science, Philosophy and History	27
Table of the Sciences	32
Philosophy of Science	36
II. THE STRUCTURE OF SCIENCE	42
Possibility of a Logic of Science	42
Method of a Logic of Science	45
Phenomenological Analysis of Science	47
Science as a Value Activity	50
The Explanation of Science	57
III. NATURE: OCCURRENTS	61
Occurrents	61
Relational Occurrents	64
Association and Dissociation	66
Space and Time	71
IV. NATURE: COMPLEXES	73
Associative Complexes	73
Dissociative Complexes	77
The Data of Science	82
The Task of Science	87
V. AWARENESS	94
Relation of Awareness to its Content	94
Intensity of Awareness	98
Kinds of Awareness	99
Locus of Awareness	102
Observation	105
Intuition	108
Thought	110
Awareness of Applicability	112

VI.	OPERATIONS	117
	Kinds of Operations	118
	Unconscious Inventions	121
	Conscious Inventions	123
	Conscious Discoveries	124
	Unconscious Discoveries	130
	Normative Aspect of Operation	134
VII.	MEANING	139
	Icons	149
	Indices	149
	General Characterizing Symbols	151
	Concepts and Propositions	159
	Correlational Symbols	162
VIII.	MEANING : CORRELATIONAL SYMBOLS	169
	Diversity	176
	Compatibility and Non-implication	177
	Co-implication, Implication, Incompatibility	186
	Contradiction	193
IX.	MEANING : CONSTRUCTS AND HYPOTHESES	198
	Origin of Suppositional Symbols	200
	Function of Suppositional Symbols	207
	Constructional and Hypothetical Methods	211
	Constructional Method	215
	Hypothetical Method	222
X.	THE DEVELOPMENT OF KNOWLEDGE	233
	Criteria of Adequacy	234
	The Development of Knowledge	238
	Pictorial Knowledge	243
	Non-Pictorial Knowledge	244
	Descriptive Knowledge	246
	Explanatory Knowledge	247
XI.	MODELS	249
	Advantages of Pictorial Symbolism	256
	Disadvantages of Pictorial Symbolism	258

XII.	Description	267
	Adequacy of Descriptive Symbols . . .	268
	Adequacy of Correlational Symbols . .	276
	Adequacy of Suppositional Symbols . .	283
XIII.	Explanation.	290
	Adequacy of Descriptive Symbols . . .	292
	Adequacy of Correlational Symbols . .	297
	Deductive Systems	302
	Theories as Probable	310
	Adequacy of Suppositional Symbols . .	312
XIV.	Quantitative Methods	316
	Integrative Value of Numbers . . .	320
	Inplications of Quantitative Explanation. .	323
Index	. . . .	329

THE LOGICAL STRUCTURE OF SCIENCE

CHAPTER I
THE PLACE OF SCIENCE

What is science?

The question is easily formulated and frequently asked. But as is the case with some of our most profound queries and some of our most recurrent difficulties, the question is asked more often than it is satisfactorily answered. The ease and frequency of its formulation lead one to expect a certain readiness and simplicity in its answer. Neither of these expectations seems to be fulfilled by the rather extensive literature upon the subject. In fact, one gets the distinct impression that many of the most significant features of science are still clouded in relative obscurity. Superficially there seems to be a rather general agreement among authorities; science uses inductive and deductive methods, employs experimental techniques where possible, claims generality, disregards questions of value, etc. But as one penetrates beneath the surface he discovers significant differences and disagreements as to the precise way in which science combines the inductive and deductive methods, the extent and limitations of experimental techniques, the range of the generality of its knowledge, the legitimacy of neglecting questions of value, etc. The conclusion seems to be that, by and large, science is not understood in any satisfactory or final sense.

As is often the case with questions which do not readily yield answers, the difficulty may be due to a lack of precision in formulation. Science, clearly, has a twofold nature. It is a discipline having a subject-matter and internal structure, but it is also a phase or aspect of human culture. Two questions are therefore suggested: (*a*) What is the structure of science? (*b*) What is the place of science in the general scheme of things? The fact that the answers to these questions may be inter-dependent does not prevent their isolation for purposes of study. The answer to the former involves such considerations as the following : the method of science, its assumptions, the validity of its results, its chief problems and the way in which they are solved, the meaning of its basic concepts and principles. The result is a synthetic view of science as the systematic whole of its elements ; one sees science as a whole because he has seen it piece-meal. But this obviously does not characterize science adequately, for science is a type of human behavior, a phase of social activity; it cannot therefore be understood apart from its logical and functional relations to the other aspects of human life. Not only is there science ; there is science and religion, science and art, science and morality, science and industry. Thus one may attempt to determine the nature of science by examining human activity as a whole and pointing out the place which science occupies in this total experience.

Moreover, the second of these two questions yields to further analysis. Two more specific questions emerge : (1) How has science in any particular historical period influenced and been influenced by the contemporary art, religion, morality and other associated human activities ? (2) What are the logical relationships which the activity we call scientific bears to those other human activities which we call art, religion, morality, etc.? The former is a question of actual causal influences, the latter is one of theoretical coordination ; the former presupposes a thorough understanding of the historical epoch, the latter involves a philosophical perspective of human activities in general. The former might be answered in terms of the modern period by pointing out how science by its emphasis on the physiology of human behavior has weakened the efficacy of moral standards in the life of the individual, how the worship of science has led to the attempt to apply its methods and criteria to the field of religion in the effort to "make religion scientific," how the recent advances in the practical aspects of science have led to a significant revision in our manner of living and in our resulting outlook upon life, how the concepts of growth and development in the biological sciences have determined a view of the universe as organic in which we must " take time seriously."

It is the task of the following pages to answer the question, What is science ? But I hope to profit by this preliminary analysis of the problem. As a result, I shall confine my attention mainly to the more specific question of the *structure* of science. The chapters which follow the present one will, in fact, be devoted to a consideration of precisely this question. But to begin immediately with this issue would be to approach the subject somewhat abruptly. A more gradual introduction seems desirable. Consequently, since one is in a better position to understand the structure of science if he sees it first in its general context, I have chosen to devote the present chapter to a discussion of the logical place of science in the general scheme of human activities and experiences. The historical, i.e. the functional, problem will be neglected completely; it is obviously of too great scope to be considered here.[1] Our problem, then, is one of determining upon the basis of a general analysis of human activities the nature of scientific activity and its logical relations to certain other disciplines of which the most important is philosophy.[2] It is the essential task of this chapter to determine what is meant by the philosophy of science.

Human Activities

The outstanding fact with reference to human activities, which may be taken as the starting-point for the discussion in this chapter, is that they are all *value* activities. The essential feature of a value activity is that it is pursued for the sake of an interest or a desire. Now the end or goal of the activity may be found in the activity itself or it may be found in something external to it. It seems to be characteristic of the basic human activities that they are pursued not for themselves but for the sake of an end which, although intimately bound up with the activity, is essentially outside of it. Thus the value activities are pursued

[1] The literature upon this problem is extensive. Typical examples are the following: E. A. Burtt, *Metaphysical Foundations of Physical Science* (New York, 1925); A. N. Whitehead, *Science and the Modern World* (New York, 1925); C. E. Ayres, *Science, the False Messiah* (Indianapolis, 1927); J. W. Krutch, *The Modern Temper* (New York, 1929); B. A. W. Russell, *The Scientific Outlook* (London, 1931).

[2] Some of the most valuable references upon this topic are the following: Robert Flint, *Philosophy as Scientia Scientiarum* (London, 1904); H. E. Bliss, *The Organization of Knowledge and the System of the Sciences* (New York, 1929); E. C. Richardson, *Classification, Theoretical and Practical* (New York, 1901); C. S. Peirce, *Collected Papers*, Vol. 1 (Harvard University, 1931), Book II; E. Goblot, *Le système des sciences* (Paris, 1922).

extrinsically for the sake of ends which are desired intrinsically. To say that the ends are desired intrinsically is simply to affirm that they are preferred in and for themselves. It seems to be impossible to get beneath the fact of preference; man, as a matter of fact, does have likes and dislikes. Furthermore he seeks that which he likes and avoids that which he dislikes. The end of the activity is the value, and the activities pursued in the attempt to gain the end are the value activities.

Now these characteristic value activities and experiences seem to fall more or less naturally into certain fairly well defined classes. In other words human experience reveals certain typical or representative kinds of activity, each of which is associated with its proper end or ideal. In presenting such a classification of human activities one is confronted with alternative methods of describing them. One may define the activities in terms of the ideals or one may define the ideals in terms of the activities. Since the two are strictly correlative the difference is of no great importance. In the following listing it will be more convenient to characterize the activity in terms of the ideal.

Perhaps the most obvious of these types of activity is that which is concerned with the satisfaction of the wants of the body; it is activities of this type which relate us most intimately with the animals. The ideal of this type of activity may be characterized as *health* or physical well being, and the essential activities will be those of eating, sleeping, exercising and resting, caring for the body in such ways as to prevent the encroachment of disease, protecting oneself from harm, etc. Closely related to this type of activity, but distinguishable from it, is that whose end is *recreation* in general without particular regard for considerations of health. Man is essentially a play-loving animal and engages in certain definite pursuits whose aim is the realization of this end. The great diversity of activities employed in the attainment of this end makes any listing of them impossible; obviously what is for one individual recreation is for another work, and what is recreation for an individual at one time may be work at another. But the fact that they are distinguishable indicates that they are different kinds of activity, and the principle of differentiation must lie in the character of the goal in the attainment of which they are employed. In such a list we find all kinds of play—individual and social, pursuit of hobbies, intellectual and physical pastimes, etc. But life is not all play, and we find that the need of man for food, clothing, and shelter gives rise to another type of activity which is usually classed as economic. The ideal in this case we may characterize as *wealth*, using this term in the ordinary sense as that which has value in exchange. Economic activities are usually defined, at least in the more highly developed societies, as those devoted to the attainment through social activity of goods which could not easily be attained through individual activity. Through co-operative enterprises man finds it possible to produce and store up goods of one kind, which may then be employed as exchange values in obtaining goods of another kind; in this class fall all activities devoted to the production and transfer of the material and immaterial goods of life. But as social beings we participate in new activities devoted to the attainment of more permanent and more satisfying relations with our fellow men. Within the life of the individual there are certain experiences which are incapable of being understood apart from what may be called values of *association*; in this class fall all of the activities concerned with the family, the community, the state and the nation. Activities related to experiences of this sort clearly fall together and constitute a type of human activity. Intimately bound up with these activities is that class which is constituted by moral activities. Here we find all of those experiences, more strictly individual than the former, which are directed to the pursuit of goodness or *character*. The generally recognized values of this type are honesty, temperance, virtue, benevolence, etc. Still more characteristically human, perhaps, are those activities concerned with the pursuit of esthetic values; man is intrinsically interested in the creation and appreciation of *beauty*. However beauty may be defined, it is possible to group together all of the activities and experiences of the individual in his attempt either to embody beauty in a material form or to extract it for his own enjoyment from the creations of nature or the productions of other beings like himself. Of somewhat the same kind are those activities called religious. These are less adequately defined though they possess enough marks in common to be grouped into the same class. Here will be found worship, prayer, devotion, faith, and allied experiences, in so far as they are concerned with the attempt to establish harmonious relations between man and the world of nature. Man finds himself in a world which is fundamentally hostile and antagonistic to the realization of many of his fondest hopes. Through propitiatory acts and intellectual constructions he attempts to allay the hostility either by bringing the world into harmony with his wishes or by explaining the fact of evil and thus seeing it as an inevitable concomitant of natural processes. The final aim of the activity is *reconciliation with nature*, and this may therefore be taken as the ideal in terms of which the activity is to be defined. Finally there are the activities which may be classified as reflective or scientific in the broadest sense of the word. In this group fall all experiences associated with the pursuit and attainment of *truth* or knowledge in its diverse forms.[1]

At this point one is tempted to loiter. Such a classification of human activities suggests a large number of interesting and important problems. But since such problems are foreign to the main content of our discussion, we may be satisfied with a mere mention of one of them.[2]

It is clear that such a list of value activities, if an ordering principle based upon preference is adopted, readily expresses one's value outlook upon life in general. For man not only engages in these various value activities but is prone to evaluate the values among themselves and to order them upon the basis of the part they play in life. This is the problem of the evaluation of values. Since conflicts often arise between two kinds of activity, it is important to have at hand a scale which will inform us as to which of the antagonistic values is more to be desired.

[1] The results of this discussion and of that which immediately follows are summarized in a table on page 32. It may be well for the reader to refer to this table in advance.
[2] It is unnecessary to point out that such a listing is largely arbitrary. Many would wish to exclude play activities, others would question the advisability of including bodily activities, still others would identify morality and religion or morality and social behavior. I note that the list as given coincides closely with the list given by W. G. Everett, *Moral Values* (London, 1920), p. 182.

For the academically minded, truth is supreme and all else is to be sacrificed to the pursuit of truth; for the religious enthusiast, values of religion are paramount; for the artist, beauty is supreme; for the materialist, wealth is the ultimate end of life. Thus such an ordering of value activities expresses one's philosophy of life, in the ordinary sense of the word. Since our concern is not with a value attitude toward the universe, no attempt at appraisal is to be read into the list presented above.

It is true that such a classification suggests that the various types of activity are more independent of one another than they are in fact. If we consider one of them, say the reflective activity, we can readily illustrate this point. It is clear, is it not, that one's science may become one's religion? If it is the task of science to understand the universe, and if by understanding the universe one can best adjust himself to its hostile features then science may become the instrument of religion very much as philosophy was at one time the handmaiden of theology. Again, it is relatively common for the scientist to be impressed by the beauty of the universe. Though this may represent an intrusion into the reflective activities, considerations of beauty, particularly of neatness and order, cannot be considered as fundamentally antagonistic to rationality. Thus art and science may be merged. Science also has its economic aspects and may be pursued for the sake of pecuniary gain; thus it becomes a type of economic activity and takes on all of the characteristics of this kind of human pursuit. Nor are moral questions entirely outside of the field of science; often it is an important moral issue whether a new discovery, say with reference to high explosives, ought to be made public. It is well to remember, therefore, that the distinguishability of these activities does not mean their independence of one another. Life is an organic unity in which these experiences merge into one another, derive their character from one another and from the situation in which they are called into play.

Among these characteristic human activities there is one which is, for our purposes, of paramount importance. This is *reflection*. It is particularly interesting because of the degree to which it has been diversified. In other words, the most significant thing about the fact of *science* is the fact of *sciences*. The various sciences are species of reflective activity, and may be discerned through the ascertainment of the proper principle of division.

The most obvious principle which suggests itself is that of *method*. Though the reflective activities are alike both in their general logical structure and in the end which is being sought, they differ in the specific methods which are being employed in the attainment of the goal. The end is one but the means are many. Though these differences are not so great as to destroy the unity of the reflective processes as a whole, they are sufficiently important to permit the determination of kinds of science through their application. The procedure in such a case would involve the selection of the methodological principle, its precise definition, and the proper grouping of the existing sciences by means of this principle. A great variety of classifications would almost certainly result. One investigator would divide the sciences into the experimental and the non-experimental. Another, basing his division upon the recognition of the importance of mathematical methods, would break up the sciences into quantitative sciences, in which the method of measurement is readily applicable, and non-quantitative sciences, in which it is not. Another, basing his division upon the more fundamental distinction of the origin of the principles constituting the science, would group the sciences into the rational or *a priori* and the empirical or *a posteriori* sciences. Still another, basing his division upon a distinction which has been much emphasized in recent philosophy of science, would subdivide the sciences into statistical sciences, where the knowledge is of the group rather than of the individual, and sciences of the individual, where the knowledge is of the individual rather than of the group.

I must confess my inability to see in such principles anything of a fundamental nature. It seems to me that considerations of method are in every case based upon considerations of subject-matter. Certain types of content permit the ready application of certain types of method. Hence the proper principle for the subdivision of reflective activities must be found in what we are attempting to know rather than in the manner in which we are attempting to know it. Methodological principles can result in nothing but fuzzy classifications. Most of the sciences are complex in method and employ a variety of types of procedure, some with great success, others with little. It is impossible to select one method and assert that this is the only, and therefore the defining, method of a particular science. Is astronomy experimental or not ? Is biology quantitative or not ? Is physics rational or empirical ? Is mechanics statistical or individual ? The difficulty lies not simply in the inadequacy of the definition but in the complex character of science itself.

The alternative approach, viz. through subject-matter, is not so simple. If we adopt this principle, the sciences will be diversified according to the qualitative diversification which the universe itself exhibits. Obviously the universe is a complex of kinds of entity, viz., time, space, number, order, motion, force, matter, life, etc. Hence the problem of the classification of the sciences is the problem of the classification of the categories. It is hardly within the scope of this book to construct a classification which makes any claim to final adequacy. Fortunately, for the purposes at hand, such a grouping may be extremely flexible. One need only examine the existent sciences, attempt to determine what their basic subject-matters are, and then fit these categories into a general scheme expressive of the way in which they seem to be organized in reality. Such a classification will itself be expressive of the state of our general knowledge at a particular period in history; in fact, the adequacy of any system of classification of the sciences must be estimated in terms of the general cultural situaton out of which it grows. Consequently the further development of knowledge may necessitate the introduction of new categories, the elimination of old categories later recognized as erroneous or redundant, the re-arrangement of categories upon the basis of newly discovered relationships, etc. All that is required of such a classification is a more or less ordered listing of the various types of entity which nature reveals. Each of the entities is an abstraction from the total complex which constitutes the universe. Thus the several sciences are *specialized* attempts to study something which we cannot in any very satisfactory way study as a whole. Such division of labor is warranted by the complexity of the universe, and the danger of being misled by abstractions is more than equalized by the possibility of concentrated effort applied to the more specialized subject-matter. Since we shall later use the word " science " in a specific sense, let us use the word " study " to designate a kind of reflective activity specifically determined by a kind of subject-matter.

The principle which I have adopted is the Comtian principle of decreasing abstractness, though in my interpretation of it the resulting classification is somewhat different from Comte's. The most abstract is the most general, and the most general is the least complex. Thus correlative principles would be those of decreasing generality, and increasing complexity. According to this principle the most abstract, the most pervasive, and the simplest of the categories will be those which are the subject-matter of metaphysical studies, or at least of that part of metaphysics which is sometimes called critical philosophy; these will include such notions as occurrent, quality, relation, time and space. This list is not meant to be exhaustive but merely a selection from a group possessing a position of fundamental logical importance in the structure of the universe. As we pass to the categories which are decreasingly abstract and general, and increasingly complex, we reach the basic categories of the mathematical studies : number, order, and quantity. On the next level come the categories of physical study in its various subdivisions : force, motion, and matter. The succeeding level is determined by the category of life, which is the subject-matter of the biological studies. Finally on the last level comes the category of mind, or human activity, which is the subject-matter of the humanistic studies.

Here again one is tempted to loiter in order to consider some of the interesting problems suggested by such a table. Having constructed a table according to the principle of decreasing abstractness, pervasiveness, and simplicity, one ventures to look for other principles which seem to be revealed in the arrangement adopted. A principle which is immediately suggested is that of the increasing dependence of subject-matter upon the preceding or more basic subject-matter : metaphysics is independent; mathematics demands the concepts of metaphysics for interpretation; physical studies demand the concepts of metaphysics and mathematics, and so on. Another principle is that of method; in this regard the classification seems to exhibit a gradual passage from the rational to the empirical, and from the highly integrated to the loosely-knit explanatory systems. Still another principle is that of a decreased interest in the type and an increased interest in the individual. Dependent on this, since the individual increases in complexity as we pass from the abstract to the concrete studies, the method of isolation becomes decreasingly applicable; this suggests the decreasing use of mathematical methods in the passage to the more concrete studies. These are a few of the principles exemplified by such a classification; many of them will be considered in greater detail in the following pages.

The next step in the classification of the reflective activities involves a further complication, though nothing new in principle. As can be readily seen the group which he have called the humanistic studies includes a large number of diverse types of investigation. The subject-matter of these investigations is the total group of human activities which are usually called mind and its manifestations. Now we have already seen that there are certain characteristic types of human activity, each of which is determined by a corresponding value judgment. Consequently for the next step in the development of our classification, viz. the subdivision of the humanistic studies, we have already at hand a guiding principle. Each of the special humanistic studies is concerned with one of the characteristic types of human activity. What this means is that man not only evaluates nature but also evaluates himself in the act of evaluating nature. Man is interested in taking pictures of nature, but he is also interested in taking pictures of the act of taking pictures of nature. In other words man finds his own value experiences interesting subject-matter for investigation. Consequently he sets about evaluating them in that particular way which we call reflecting about them; in other words, he evaluates them from the point of view of that value which is called truth. This act involves a peculiar turning-inside-out of the table of reflective activities—a fact which results in no way from an inadequacy of the diagram but is a necessary character due to the nature of the world which the diagram is attempting to portray.

The actual humanistic studies are then somewhat as follows : *studies of health*, e.g. medicine, physiology, the so-called science of public health, and all related studies ; *studies of play*, e.g. physical education as commonly taught in universities today, psychology of play, and cognate studies ; *studies of working activities*, e.g. economics ; *studies of social activities*, e.g. sociology, political science, science of law, education ; *studies of moral behavior*, e.g. ethics ; *studies of art*, e.g. esthetics ; *studies of religious activities*, e.g. science or philosophy of religion ; and, finally, *studies of reflective activities*, e.g. science of science.

Our analysis might well stop at this point. But the trend of the development in recent years seems to indicate that one of the humanistic studies, viz. studies of reflection, is more complex than has been commonly recognized. It contains, in fact, certain well defined subdivisions, which are taking form and structure, acquiring a literature, and establishing themselves as legitimate intellectual enterprises. The principle according to which such disciplines should be established is found within the classification as revealed up to this point. We have attempted to show that an intellectual discipline arises whenever reality reveals a definite type of entity, concerning which generalizations may be asserted. Now the specific studies which we have already listed are certainly such definite types of entity, possessing individuality among themselves and exhibiting structural characteristics. Consequently a discipline which studies reflective activity is bound to recognize the necessity for studying the specific reflective activities exhibited by the specific disciplines. And the study of each of these studies may in turn become an independent intellectual discipline. Thus there will be studies of metaphysical studies, studies of mathematical studies, studies of physical studies, studies of biological studies, and studies of humanistic studies. Following out the same principle, since humanistic studies are of many kinds, there will appear to be studies of medical studies, studies of play studies, and so on, down through studies of religious studies, and ending with studies of studies of reflection. There is reason to believe that we have here reached a discipline with a sufficient degree of remoteness from the concrete studies of every day life to satisfy the most ardent lover of abstractions; to study about studying is quite difficult enough for most minds, but the discipline which we just mentioned is a study of studying about studying, i.e. a study of the methods of the logician while studying the methods of the scientist. It should be noted, however, that there is nothing in principle which demands that we stop here.

Science, Philosophy, and History

We now pass to a very important point. Throughout the discussion thus far we have carefully avoided the use of the words, *science, philosophy,* and *history,* except where we employed them for illustrative purposes. We are now ready to give these terms a precise significance. They designate three kinds of study, or three kinds of approach to the intellectual grasp of reality[1]. Yet they are not distinguished primarily upon the basis of method; here again it seems advisable to make considerations of method secondary and considerations of subject-matter primary. It is because the world exhibits three aspects that we have these three approaches. The world is temporal. Being temporal, it permits us, first, to describe its succession of changes, i.e. to write its history; it permits us, secondly, to isolate cross sections, i.e. to portray its science; it permits us, thirdly, to abstract its enduring features, i.e. to reveal its philosophy. Thus we create three distinct disciplines when we attend to these three aspects of nature. Of anything in nature we may ask three questions: How did it come to be as it is? What is it and what are its present relations to other things? What of its features have remained constant throughout its development? The answers to these questions tell us its origin, its present features, and its enduring nature. The historian sees his subject matter in its aspect of *becoming*; the scientist and the philosopher see it in its aspect of *being*;—but the scientist sees

---

[1] This distinction is recognized by Bliss, *op. cit.*, pp. 302–3. To the three here given Bliss adds a fourth, viz. applied science.

it under its *temporal being*, while the philosopher sees it under its *eternal being*. Thus, illustrating by means of the categories, we may say that the historian is interested in the development of the categories, the scientist makes an analysis of them at the relatively low level of generality which is their manifestation at a given period, and the philosopher generalizes from science and history and determines the categories at a level of generality which will include all of their historical manifestations.

It follows that our table of reflective activities must be reproduced in triplicate. For each of the studies represents the possibility of a science, a history, or a philosophy. Thus, for example, there will be scientific biology, historical biology, and philosophical biology; and within the humanistic studies there will be scientific ethics, historical ethics, and philosophical ethics. Furthermore, the possibility of disciplines which study disciplines results in a history of science, a history of philosophy, and a history of history, as well as in a science of history, a science of philosophy, and a science of science. In fact there may even be such disciplines as the philosophy of the history of science, or the history of the history of philosophy. While some of these combinations are theoretical possibilities only, and not exemplified in any cultural set-up, the revelation of their abstract possibility is enlightening.

Although the theoretical distinctions between these three types of study are relatively sharp, the actual designation of a given study as a science, a philosophy, or a history cannot always be made with certainty. This is due to two facts: (*a*) The highly abstract features of the universe do not seem to exhibit changes in their exemplifications through the passage of time; hence the studies of high abstractions exhibit an absorption of science and history into philosophy. (*b*) The human mind is dissatisfied with relative concretions; hence the sciences and histories of the relative concretions tend to pass into the philosophies of those concretions, and the philosophies of relative concretions tend to pass into the philosophies of high abstractions. Let us consider each of these points.

(*a*) Since the basic categories do not exhibit development, the studies of these high abstractions will be essentially philosophical rather than scientific or historical. This is clearly the case in those studies which are called metaphysical. It is impossible to distinguish between scientific metaphysics, historical metaphysics, and philosophical metaphysics simply because such notions as occurrent, quality, relation, time, and space do not manifest themselves in different forms in time. These categories represent, rather, the general structure within which change must take place. Though there may be a philosophy of time there does not seem to be significantly either a science or a history of time which can be distinguished from it. The same merging of disciplines seems to apply to the concepts of number, order, and quantity. Certainly there is no historical mathematics —which does not mean, of course, that there is no history *of* mathematics. Although a distinction has arisen in recent years between mathematical science and mathematical philosophy, the principle employed in this division is that of logical generality rather than that of temporal permanence. But in either event the lines of demarkation between the two disciplines can be drawn only with difficulty. In physical studies, however, the historical aspect emerges not only as a conceivable study but as an actual discipline already in possesson of established methods and content; cosmology, cosmogony, and historical geology are instances of such studies. To this extent there is also a distinction between the science of matter and the philosophy of matter. Science attempts to determine what matter is; philosophy attempts to determine its constant feature as exhibited by what it has been and is. The distinction between the three disciplines becomes sharper as one passes through the biological and humanistic sciences. Consequently the essential difficulties in separating one discipline from another occur only upon the level of the higher abstractions.

(*b*) It is also hard to distinguish the three studies one from another because the tendency in the development of both science and history is toward higher abstractions, i.e. toward philosophy. Thus since the transition is unbroken, any line which may be drawn will be somewhat arbitrary. The continuity of this transition is illustrated in the fact that history is not satisfied with a mere delineation of the succession of events;

it demands also the revelation of the *laws* of the succession. But the laws represent the permanent and enduring aspect, i.e. the philosophical subject-matter. Thus every history tends to become a philosphy of history, and one cannot say when it ceases to be one and becomes the other. So also the continuity between science and philosophy is unbroken, for science ever proceeds in the direction of higher abstractions. This is exhibited in two ways. In the first place, every science of a given subject-matter tends to become a philosphy of that subject-matter through an increased emphasis on the permanent and enduring features. Thus biological science tends to pass into biological philosphy as one continues to search for the defining properties of life, i.e. those minimal properties which must be exhibited by any living organism at any time. So also humanistic science tends to pass into humanistic philosophy as one continues to search for the permanent features of human nature. More specifically a descriptive ethics becomes a philosphical ethics, a phenomenological study of art tends to pass into an analysis of art, and a study of a particular religious culture inevitably develops into an abstract study of the permanent features of religion in so far as they are illustrated in this specific situation. But in the second place every philosophy of a given subject-matter tends to become the philosophy of a more highly abstract subject-matter. Since the high abstractions possess a greater permanence, the greater concretions may be made to share this permanence through the establishment of correlations. Thus since life exhibits a greater permanence than human nature, we try to find the permanence of human nature in the organism with which it is always associated. In the same way, since matter exhibits a greater permanence than life, we try to find the permanence of life in matter with which it is always associated. The same tendency can be seen in the " reduction " of secondary qualities to primary qualities, and of primary qualities to their measured values ; it appears that quantitative features are more permanent and enduring than qualitative features. Thus every science tends to become a philosophy and every philosophy of a relatively high concretion tends to become a philosophy of a relatively high abstraction. The passage from either a history or a science into a philosophy cannot be readily ascertained, since there is no change in subject-matter; the only change is an increase in the generality of the subject matter. But the passage from a philosphy of a concretion into the philosophy of an abstraction can usually be detected, since there is always involved the introduction of a new concept.

Thus we are compelled to recognize that although there are three kinds of discipline we may not be able to state unequivocally in any given case whether a study is science, history, or philosophy. In practically all cases it will be a combination of all three. Usually it will emphasize one to the exclusion of the others. For this reason it seems best so to construct the table of disciplines as to be wholly independent of this distinction. We must then recognize that every time the word " studies " occurs in the table it is susceptible of a triple interpretation. This will permit such things as a history of a history of history, in which the forms of study are all alike, and a history of the philosophy of science, in which the forms of study are all different. At least these will be disciplines which will be abstractly possible.

Before turning to the table we must clear up one source of ambiguity. The word "of" has an unavoidable duplicity in meaning in such a phrase as the " science of mathematics." It may mean " the science which is mathematics " or " the science which studies the mathematicians at work." The ambiguity is clearly evident in the phrases " science of physics " and " science of matter." We may avoid this difficulty by employing the word only in the latter sense, i.e. as designating the relation of a discipline to its subject-matter. Then the expression " science of mathematics " will designate that discipline which studies mathematical methods and will never be used as synonymous with the science of number, order, and quantity.

TABLE I

Table of the Sciences

5. Studies of Humanistic studies	5. Humanistic studies	5. Human activities		Ideal
a. Studies of medical studies	a. Medical studies	a. Bodily		a. Health
b. Studies of play studies	b. Play studies	b. Play		b. Recreation
c. Studies of economic studies	c. Economic studies	c. Work		c. Wealth
d. Studies of social studies	d. Social studies	d. Social		d. Association
e. Studies of moral studies	e. Moral studies	e. Moral		e. Character
f. Studies of artistic studies	f. Artistic studies	f. Art		f. Beauty
g. Studies of religious studies	g. Religious studies	g. Religious		g. Reconciliation
h. Studies of studies of reflection	h. Studies of reflection	h. Reflective	Subject-matter	h. Truth
	1. Studies of meta. studies	1. Meta. studies	1. Occurrent, qual., rel.	,,
	2. Studies of math. studies	2. Math. studies	2. Number, order, quan.	,,
	3. Studies of phy. studies	3. Phy. studies	3. Force, motion, matter	,,
	4. Studies of biol. studies	4. Biol. studies	4. Life	,,
	5. Studies of human. studies	5. Human. Studies	5. Human activity	,,

Some comments are necessary in explanation of the table. In order to simplify this explanation let us adopt a concrete interpretation of the table in terms of sciences rather than in terms of histories or philosophies.

The table is devised for the purpose of indicating in a schematic form the interrelations of the various disciplines of which the sciences are one particular type. It thus aims to represent the place of science in the general classification of intellectual activities. In order to do this successfully it must show both how the various sciences are differentiated from one another and how science as a whole constitutes a kind of human activity having specific relations to other kinds of activity. The table indicates, if it is properly understood, that the humanistic, biological, physical, mathematical, and metaphysical sciences, are all kinds of that characteristically human activity which is called science in general, and that science in general is itself one among various human activities. But it also indicates that the subject-matter of the humanistic sciences is this very field of human activities of which science is one. I have tried to represent this fact by making the columns which end with the number 5 at the bottom continue in the next column at the left starting from the top. Thus " human activity " at the lower right is identical with " human activity " at the upper center; so also " humanistic studies " at the lower center is identical with " humanistic studies " at the upper left. The subject-matter of each study is indicated in the column immediately to the right. Since the human activities themselves do not have any subject-matter the column representing the various *ideals* of the activities takes its place.

The diagram aims to represent, in the second place, the difference between descriptive and normative sciences. According to the commonly accepted distinction, the former are those sciences which merely symbolize a certain subject-matter while the latter are those which not only describe but also evaluate their subject-matter according to a norm or standard. It is clear from the diagram that metaphysical, mathematical, physical, and biological studies are descriptive; they attempt merely to portray the character of certain important aspects of being. But

B

the humanistic studies are all normative, for they not only describe the various human activities but also examine the correlated ideals in terms of which the efficacy of the activities is to be estimated; the ideal is in each case a part of the subject-matter. Thus medicine, for example, is not merely a study of the way in which our bodies function; it is also a consideration of the way in which they should function if we are to attain health. Again, ethics is not simply a study of the way in which we do behave in problematic situations; it is also a study of the way in which we should behave in the interests of attaining goodness of character. The fact that the ideal of truth is necessarily involved in all reflective activities might easily lead one to believe that *all* sciences are normative. But this indicates a confusion. In the case of the metaphysical, mathematical, physical, and biological studies truth is not part of the subject-matter of the investigation; it is rather the principle which guides the investigation as a whole. The economist must define wealth, but the astronomer need not define truth.

The table indicates, in the third place, how some of the so-called conflicts between certain studies and certain kinds of activity are to be resolved. Consider for purposes of illustration the supposed hostility between science and religion. This antagonism arises through an ambiguity in either or both of the terms, and can be resolved only by avoiding this equivocation. By religion we may mean religious activities, i.e. the behavior and experiences of an individual in the attempt to reconcile himself with nature. By science we may mean the totality of reflective activities or any one such activity. In either case there can be no serious conflict, for science is then another type of activity in search of another ideal. The conflicts will be such as may occur between work and play, or art and morality— difficulties which must be met by ascertaining which of the rival values is to be considered as paramount in the situation at hand. But by religion we may mean the science of religion, i.e. the science which studies religion. Even so, there can then be no conflicts more serious than those which often arise when respective fields of study are not sharply defined as to subject-matter. In general the sciences are not in conflict; they are united in a cooperative enterprise. Again, by religion we may mean a religious creed, i.e. the theory of the nature and origin of the world which is derived through the religious experiences. Here there may be a conflict with science, for it is the task of the latter to determine the character of the world ; when religion takes over this task, it is usurping the field of science. Thus such a table as this serves, if not to solve the problem, at least to locate the difficulty and thus to determine the general nature of the solution.

In the fourth place, certain omissions from the table require explanation. Notable among these is psychology. But a closer examination will reveal its presence in the only way in which is can be conveniently shown in the diagram. In the broadest possible sense of the term, psychology is identical with humanistic science in general ; there are then as many subdivisions of psychology as there are distinct kinds of humanistic science. Thus there is a psychology of health, of work, of morality, of religion, and, in this sense, anything which the physician or the economist or the ethicist or the scientist of religion discovers will be a psychological fact. But in the narrower sense psychology has considered itself to be concerned with certain abstractions from human activity rather than with the totality of human behaviour. On the one hand, these abstractions lead to a consideration of bodily activity in dissociation from the conscious correlates of that activity; this results in behavioristic psychology. On the other hand, the emphasis may be upon the conscious correlate, not as it manifests itself in the differing types of human activity, but as it constitutes the general pattern or form of all human experiences. In this sense psychology is an attempt to define consciousness as such, in view of its diverse manifestations in the various forms of human activity and experience. The limited dimensions of the table do not make possible the representation of these kinds of psychology. But psychology in its most general sense is clearly represented, though it is not so named in the diagram. Certain composite sciences have been omitted simply because they can be analysed into several of the more basic sciences. Their function is the revelation of the peculiar interrelations of the sciences to form complexes not capable of direct representation in the diagram. For example, geography, which is a study of man in relation to his environment, is clearly a composite of certain aspects of geology, biology, economics, sociology, science of religion, and perhaps other sciences. Many of the practical sciences, e.g. navigation, engineering, ceramics, aviation, social administration, crime prevention, have positions which cannot be clearly indicated in the diagram. But the difficulty is due only to the complex character of the enterprises, and their omission from the table is not to be taken as implying that they are not genuine cultural aspects of the modern world.

Philosophy of Science

For our purposes the most important of the studies is that which has been called in recent years the philosophy of science. The very fact of such a discipline suggests that the old hostility between sciences and philosophy is disappearing. There is even some reason to believe that they are becoming united with sufficient harmony to permit the definiton of one in terms of the other; Flint, for example, defines philosophy as the " science of the sciences."[1] Let us examine our table with a view to determining the scope of such a discipline as the philosophy of science.

But first let us glance at the table for the purpose of discovering what place or places philosophy as ordinarily understood occupies. (*a*) Philosophy includes all metaphysical studies as the analysis of the basic categories. This is sometimes called analytic metaphysics or critical philosophy, though it is not quite identical with what Broad[2] describes by the latter term. As thus defined, metaphysics involves no overlapping with the fields of the special sciences, for the scientist as such does not concern himself with these problems ; if he does feel himself under obligation to consider them he is clearly to that extent a metaphysician. From this point of view metaphysics could be defined as that reflective method which examines the basic categories in the same way that any other science examines the less basic concepts.

[1] Flint, *op. cit.*, pp. 3–63.
[2] C. D. Broad, *Scientific Thought* (London, 1923), p. 18.

We have already seen what there is of legitimacy in this definition. Metaphysical studies constitute a genuine type of reflective activity. But such studies are to be called philosophical rather than scientific or historical merely because the science and the history are absorbed into the philosophy. In order to indicate clearly the relation between philosophy and science as ordinarily understood, let us define this aspect of philosophy as *the study of the categories implicit in the sciences*.

(*b*) Traditionally philosophy has been defined in such a way as to include that group of humanistic studies which we have designated ethics, esthetics, studies of religion, and studies of reflection. These are generally recognized to be legitimately included in philosophy; whereas their close neighbors, sociology, economics, etc., are commonly excluded from philosophy and called sciences.[1] There seems to be no valid reason for this sharp distinction. Certainly the investigations are of essentially the same kind, and the subject-matters are not fundamentally different. The greater specialization which is evident in recent philosophy would seem to indicate that the investigations into moral behavior, art, religion, and reflection can be carried on more or less independently of one another. Thus the division of labor among the philosophers is quite the same as that among the scientists; hence the fact that the ethicist and the esthetician choose to call themselves philosophers rather than scientists is of no particular significance. Perhaps one might argue that morality, art, religion, and reflection are the more distinctly human activities while the other activities are more or less completely shared by the lower animals; then the investigations into such activities would be essentially different from the investigation into the less representative activities. But such a distinction could have nothing more than a practical working value and is almost certain to be obliterated with the advancement of our knowledge of animal behavior and with an increased recognition of the absence of a sharp line of demarkation between man and the higher animals. If ethics, esthetics, and the science of religion could be shown to be

[1] Though it is perhaps not generally recognized that they are even to be called *sciences*.

studies of intellectual disciplines, then there would be reason for calling them philosophies in the special sense in which the word is sometimes used to designate a discipline which underlies another; thus the discipline which studies a discipline is commonly called a philosophy of that discipline. But this is hardly a justification for calling *all* these studies philosophies; morality, art, and religion are not intellectual disciplines in the same sense that reflection proper is; they are essentially ways of behaving which exhibit more or less predominantly an intellectual element. Thus there seems to be no satisfactory reason for calling ethics, esthetics, and studies of religion philosophies in any special sense. They are philosophies only in the sense in which we have already considered them to be, viz. each of them is a science, a history or a philosophy depending upon the way in which the subject-matter is studied.

But the last of these studies, viz. the study of reflection, deserves special consideration. It is to this discipline that the term "logic of science" as a part of a general philosophy of science has been applied. However it is named, it has a well defined subject matter. It is a study of the activity which is science itself, and involves a consideration of the method of the science, its logical structure, its assumed theory of knowledge, the operational techniques employed in defining its concepts, etc. The remaining chapters of this book will be devoted to the determination of some of the conclusions in this field. The discipline may be broken up into sub-disciplines, each of which may endeavor to establish certain results of narrower scope. There is reason to believe that such specialization is taking place to-day. As a result there arises a logic of metaphysical studies (which actually does not exist as an independent discipline, since the metaphysician usually studies his subject-matter self-critically, i.e. he examines his own method), the logic of mathematical studies, the logic of physical studies, the logic of biological studies, and the logic of humanistic studies. There are probably some even more specialized studies, e.g. the logic of economics or the logic of sociology, though these have hardly emerged as distinct studies. Philosophy in this sense may be defined as *the study of the methods of the sciences*.

(*c*) But philosophy as ordinarily understood occupies a third place in the table—a place which cannot be clearly indicated. The specialized nature of the investigations of the scientists creates a new problem; this is the problem of the coordination and interrelation of the results of the sciences with a view to integrating them into a totality. The systematic attempt to solve this problem is usually called speculative philosphy.[1] It has always been the task of the philospher to see things in their wholeness. The speculative philosopher is thus seeking a *Weltanschauung*. His task is the consideration of the problems of the inter-dependence of the fields of the special sciences, and his results will therefore depend upon an acquaintance with the results of critical philosophy. The characterization of this type of philosophy as speculative should not be looked upon as derogatory; although its results can never be as certain as those in the critical field, many conclusions can be derived which possess a high degree of certainty. The difficulty in representing this type of philosophy in the diagram lies in the fact that the diagram is itself an instance of it; i.e. one of the problems of speculative philosophy is precisely that of the classification of the sciences. Hence one could perhaps indicate the place of speculative philosophy in such a table by writing it underneath as the title of the diagram itself. Let us define philosophy in this sense as *the study of the interrelations of the sciences*.

This much we can learn of philosophy from the traditional point of view. But this is not all of philosophy, as we have already seen. Philosophy is also one of the three fundamental approaches to the world; thus there is a philosophy correlative with each science and each history. The task of such a study is the investigation of the concepts designating the basic subject-matter of each of the disciplines with a view to revealing its permanent and enduring features. The result is a series of philosophies: philosophical metaphysics (which does not exist as distinct from other metaphysical studies), philosophical mathematics, philosophical physics, philosophical biology, and philosophical humanistic study, which includes

[1] Cf. Broad, *op. cit.*, p. 20; M. Cohen, *Reason and Nature* (New York, 1931), pp. 147–8; Whitehead, *Concept of Nature* (Cambridge, 1920), p. 1.

philosophical economics, philosophical ethics, and so on. It is the fact of a philosophical ethics which accounts for the usual placing of ethics in the class of philosophical rather than scientific disciplines ; although there is a scientific ethics, it is lost sight of in the philosophical ethics. So also with esthetics and the philosophy of religion. This part of philosophy may be defined as *the study of the categories explicit in the sciences*.

We may now summarise our results as a preliminary to the consideration of the meaning of the philosophy of science. Philosophy has been defined as a study (*a*) of the categories implicit in the sciences, (*b*) of the interrelations of the sciences, (*c*) of the methods of the sciences, and (*d*) of the categories explicit in the sciences. Such a definition as this indicates clearly the relation between philosophy and science, and readily permits the definition of philosophy as the science of science. If one wishes, therefore, to be as generous as possible he may define the philosophy of science as identical with the whole of philosophy ; there is clearly a sense in which all philosophical problems are connected in the variety of ways suggested above with some feature of science. But the field of the philosophy of science would not gain in precision by being thus enlarged in extent. It seems advisable, consequently, to limit the field somewhat more narrowly.

The philosophy of science, in this narrow sense, consists of two parts : the study of the methods of the sciences, and the study of the categories explicit in the sciences. These are commonly called, respectively, the logic of science and the metaphysics of science[1] (if one recalls that it is metaphysical only in that it is studying categories and not because it is studying the all-pervasive categories). The interrelation of these two parts is obvious. The philosopher of science attempts to determine the meaning of the basic concepts of the science in question with distinct reference to the method which has been employed in the determination of those concepts. He will attempt to determine how the concepts are derived from experience, how

---

[1] Clearly the " of " in " metaphysics of science " shares the ambiguity which was noted earlier in the chapter ; it is not really a metaphysics which studies science but a metaphysical study which is associated with a scientific study.

their contents are determined by their manner of derivation, and how they are often redefined in order to accomplish a higher degree of integration in the conceptual system of which they are parts. Thus the philosopher of science must analyze the world to determine its features, and science to determine its methods. And he must ascertain whether the methods are adapted to the subject-matter and to what extent the knowledge of the subject-matter is a reflection of the methods. The philosophy of any one science, e.g. the philosophy of mathematics, will consist of these two parts : a logic of mathematics and a metaphysics of mathematics. So also there will be a logic of biology and a metaphysics of biology, a logic of ethics and a metaphysics of ethics, and so on.

It is true that the philosophy of science, in this sense, occupies a peculiar position among the sciences ; this may account for its being called the philosophy of science rather than the science of science. The philosophy of science is to a certain extent presupposed by all of the special sciences. For Aristotle logic was a bridge to the special sciences.[1] There is reason to question the view that logic is temporally prior to the sciences ; certainly many scientists do not study logic before engaging in scientific pursuits. But there is a dependence of science upon the philosophy of science in the sense that the scientist must go to the philosopher, or must himself become a philosopher, for the justification of his method and for the clear understanding of the meaning of his basic concepts. Since method determines results, considerations of method cannot be neglected. Recent quantum physics is learning this fact to its own confusion. And since the understanding of a concept always implies a higher abstraction in terms of which it can be explained, considerations of the philosophical character of the basic concepts cannot be neglected. Recent relativity physics is learning this fact, again to its own confusion. The philosophy of science represents the attempt to study deliberately and systematically the very problems which the scientist is driven to consider in the course of his pursuits. It is the acknowledgment that a study which is inevitable must be recognized as legitimate and made as precise as the character of the subject-matter permits.

[1] W. D. Ross, *Aristotle* (London, 1923), p. 20.

CHAPTER II

THE STRUCTURE OF SCIENCE

We now pass from speculative philosophy to logic. It will be the task of this chapter to make an analysis of reflection. Reflection, we have seen, is one among many value pursuits, especially interesting because of its high diversification. Our problem is to determine the general character and structure of those reflective processes which by virtue of having been subjected to a certain amount of examination and control are usually called " critical " as opposed to " common sense " or " popular." Though the discussion in this chapter will be about *science*, the generality of the considerations here advanced will make the conclusions equally applicable to *history* and *philosophy*. We are interested, in other words, only in those general features of critical reflection through which it may be distinguished from other closely similar types of value experience.

But there are two very important preliminary questions: (*a*) Is it possible to make a scientific study of science? (*b*) Granting that such a study is possible, what must be the method employed? Let us attempt to answer each of these questions.

Possibility of a Logic of Science

The essential difficulty here, as can be readily seen, is the possibility of employing a certain method in a study of something which is precisely that method itself. Method and subject-matter are always interrelated in a peculiar way; what one ascertains about a subject-matter is a reflection of the method which has been employed in arriving at his conclusion. Thus it appears that one ought to make a critical examination of one's method before he uses it to determine conclusions with reference to any subject-matter. But in order to make such a critical examination he must employ either the very method whose legitimacy he has just called into question, or some other method the legitimacy of whose use immediately becomes questionable upon the same grounds. Thus one's knowledge of a subject-matter depends upon one's method, but one's method is itself a subject-matter which must be critically examined.

Presumably the best way to meet this difficulty would be to insist that it is not genuine. The test of the legitimacy of making a scientific study of science lies not in the preliminary analysis and justification of the method but in the effectiveness of the results. To prove that science may be known, one need only plunge into the study of science and by learning something of its character demonstrate the futility of having asked the question. But would this be a conclusive demonstration ? There have been astrologies, alchemies, theosophies and other forms of quackery ; may not science lie in the same category ? The fact that there are studies which have been called science does not establish the legitimacy of science as a method of knowing. Furthermore, since we identify science essentially by its method, how can we be sure that the discipline which we are studying really is science ? Until one knows the significant features of the scientific method, he cannot locate science at all.

So far as I can see, there is no way of meeting these difficulties except by assumption. But it is well to be clear as to precisely what is involved in making such assumptions. They seem to be two in number : (1) We assume that there is such a thing as science, which possesses certain identifiable features and can thus be recognized. (2) We assume that an examination of method need not be prior to the use of that method.

(1) That science is a genuine rather than a pseudo-object cannot be conclusively demonstrated. The only final proof of reality is the denotative gesture. But this fails in the case of all value activities, for the identifying feature in such situations is a subjective factor to which one cannot point. Presumably a scientist working in a laboratory would constitute an " instance " of science to which one could point. But there is much that goes on in a laboratory that is not science. The essential feature of science is not instrumental set-ups, laboratory techniques, and the like, but the " mental " processes guiding and directing all such acts of behavior. But we can be sure that such processes are present only in terms of their " results;" hence the denotative gesture fails. It is true also that there is no proof of the unreality of science. It is the fate of all pseudo-objects to die a lingering death ; we must wait indefinitely for the denotative gesture

which will enable us to locate them. The case is somewhat better for the science of science than it is for astrology. For the denial of science implies its admission. To assert that science does not exist is to make a scientific affirmation; hence science does exist. To deny that there is such a thing as knowledge is the implicit assertion of its reality, for one must know that there is no knowledge; if there were no knowledge we could not know it. This implies that a thorough-going agnosticism and a complete skepticism are impossible, for they are both self-refuting. The simplest way to avoid all of these difficulties is to assume openly, at least as the starting-point for study, that there is something which at least claims to be science. It may prove later to have inferior value as an instrument on the claims which it makes, but we can discover this only by attempting to learn something about it.

It therefore seems reasonable to assume that science has certain distinguishing features which will enable us to identify it and distinguish it from closely related activities. We have already seen that it may be distinguished from other value activities. But this does not mean that it may be precisely defined; no field of investigation is definable prior to the investigation itself. Thus in our attempt to study science we may find ourselves unwittingly excluding many situations which are really scientific and including many situations not essentially scientific in character. This risk must be run. But it is a risk only if one supposes that a field once defined cannot be redefined. It is characteristic of all systematic investigations that the fields are being continually realigned in the course of the study, and the fact of such fluctuation does not argue for the impossibility of defining fields of study. Consequently whatever may be the difficulties in defining science precisely, there is no reason to suppose that it is lacking in such properties and hence is an illegitimate object of study.

(2) That knowledge may be *legitimately* known is also incapable of demonstration. The difficulty here is theoretically insurmountable, since the instrument of the investigation is identical with the subject-matter and there can therefore be no

preliminary critique of method. But might there not be a *concomitant* critique of method? For example, might we not avoid the difficulty by setting two logicians to the task of studying method? Then each logician could study and estimate, *pari passu*, the method of the other, and each could then confer with the other for the justification of his results. Fortunately the situation is not so intricate as this. It is possible to estimate the validity of a method *posterior* to its use. The logician can proceed in a cautious, self-critical way in the study of science. This involves as much attention to methods while they are being employed as is possible, with a continual checking of the results of one's study over against the methods which have been used up to that point in establishing these results. One then uses, so to speak, a minimum of method until he has learned something about method; he then employs only such method as he has already studied. But he then re-examines the total structure after the completion of the enterprise in order to be sure that the method of its establishment is consistent with its results. He thus, in a sense, has tried to know knowledge and has met with a certain degree of success.

Method of a Logic of Science

Granting the possibility of trying to know science, we must now endeavor to ascertain specifically how we must proceed in such a study. Here our difficulty is, so to speak, upon a different level. It is my fundamental belief that the method to be employed in the study of science is precisely the method of science itself, and it is the task of the entire book to present an analysis of this method. But we find ourselves here confronted with a situation in which we must employ the method. Specifically our difficulty is that we cannot write chapter II until we have written the rest of the chapters, yet chapter II represents that preliminary analysis of the general structure of cognition without which the rest of the chapters cannot be written. So far as I can see, the only way to meet this difficulty is to proceed with the study of the present chapter and leave the justification of its methods for the later pages.

However, it will perhaps not destroy the continuity of the discussion too much if we point out the essential features of that method so far as the present analysis is concerned. We shall first have to determine what seem to be the indubitable elements of scientific cognition. But here an important distinction arises. By science we may mean that essential core of the total reflective activity which may be expressed simply as " the awareness of a content," or we may mean the larger situation which includes also a recognition of the grounds for the awareness, the criteria for its justification and the modifiability of its nature in the direction of greater perfection. But the group of data will be different in the two cases; specifically, the data of the latter will include those of the former and will contain additional elements. We must then decide which of these two groups of data is to be taken as the empirically revealed elements of cognition. We shall find reasons for deciding in favor of the more inclusive group.

But we shall then see that science is not content with mere description. If we insist, as I shall in the later chapters, that description and explanation are strictly continuous and that science passes imperceptibly and inevitably from the former into the latter, then to stop the analysis at this point is somewhat arbitrary. Science is not merely that which is given indubitably when we confront ourselves with that which we call science; it is also all of the hypothetical and constructual elements which are required in order to make that given intelligible. Two of these, in particular, will concern us, viz. the empirical character of symbols and the rationality of nature. The former will be discussed very briefly, since it is precisely the topic to which the body of the present book is devoted. The latter will also be discussed, but again very briefly, since it is a topic which is essentially irrelevant to the question of the logical structure of science. Our complete discussion of scientific cognition in this chapter will therefore contain the following steps: (*a*) the phenomenological analysis of science, (*b*) science as a type of value activity, and (*c*) the explanaton of science.

Phenomenological Analysis of Science

In the most general terms possible, scientific cognition may be described as an occurrent.[1] But it is an occurrent which is complex and therefore contains sub-occurrents, which are its aspects. These aspects are not distinguishable from one another or from the total cognitive occurrent by spatial or temporal features; cognition is the spatio-temporal togetherness of these aspects, and the aspects can be distinguished only through their qualitative features. Thus we may say that scientific cognition is the occurrence in a given spation-temporal situation of a complex consisting of certain sub-occurrents in describable structural relations to one another.

Proceeding analytically, we discover first that scientific cognition exhibits two main aspects. These may be called the *awareness* and the *content of the awareness*. The essential feature of these two aspects is that they never occur in isolation, i.e. awareness is never found apart from content, and content is never found apart from awareness. The awareness possesses an irreducible relation to the associated content which may be expressed by saying that the awareness is *of* the content. The awareness may be inaccurately described as a feeling of tenseness, sometimes vaguely localized in one of the sense organs but usually not localized at all. It is subject to quantitative variation which permits us to say that one awareness is more intense or clearer than another. The content may change in the presence of the awareness, and the awareness may thus change its locus. A change in locus is frequently described as a change in kind of awareness, i.e. kinds of content determine kinds of associated awarenesses. For example, we imagine certain contents, remember other contents, perceive still other contents. Awareness seems to be under control from these points of view; i.e. we are able to vary its clarity, locus, or kind within certain limitations. Awareness often seems not to be an element of cognition, for the act of thinking reveals itself as a succession of contents. But by sharply attending to the contents we can see that each has

[1] This term will be explained at length in the next chapter. For the present the meaning will be sufficiently clear.

its associated awareness.[1] Thus the awareness is empirically revealed as an element in the total cognitive situation.

But there is recognizable also a very important distinction between kinds of content. This is the distinction between symbolic and non-symbolic contents. For present purposes we may conveniently define symbolic contents, and then define non-symbolic contents as those contents lacking symbolic character. The characteristic feature of a symbolic content is its meaning; thus we may assert that a symbolic content is any content whatsoever which has meaning. At this point we may simply say that the meaning property of a symbol is that by virtue of which the content refers to something else or is directed away from itself. Though a symbolic content points in a definite direction, the object of the pointing may not be given in the content or as an additional content. But at the same time the object is given, in a sense, in the pointing toward it. Thus we may know what a symbol means without being aware of the entity which is meant; we say that we are aware of the reference without being aware of the referent. The commonest example of a symbolic content is language, though any content whatsoever may conceivably be symbolic. Symbols mean in different ways depending on the type of symbol. Thus whenever any content possesses this unique property of meaning, it may be characterized as a symbolic content.

A non-symbolic content, on the other hand, is simply any content lacking the property of meaning. Such contents are the ordinary "objects of knowledge," i.e. the things which are known. The awareness of such non-symbolic contents may *always* be associated with the awareness of symbolic contents which mean them; certainly it is *usually* so associated. This fact, however, is somewhat irrelevant. The important thing to recognize is that scientific cognition does reveal a non-symbolic content as a distinguishable element. This content is simply an occurrent as occurrent, i.e. as a bare happening possessing the form and content of an occurrent.[2] Such contents are found in

[1] The reality of this aspect of the cognitive situation has been emphasized by G E. Moore, in his *Philosophical Studies* (London, 1922), chap. I, and by S. Alexander in his *Space, Time, Deity* (London, 1920), Vol. I, p. 11.
[2] Cf. chap. III, p. 63.

every cognitive situation both in the movement of symbolic discovery, i.e., the stage of deriving symbols from occurrents, and in the movement of symbolic verification, i.e., the stage of checking up of derived symbols by reference to given objects. In other words, it is through the fact of non-symbolic contents that we determine the meanings of our symbols and verify their applicability. A symbolic content may become non-symbolic either by the neglect of its meaning or by the consideration it as the referent of another symbolic content.

Finally, there is discernible in the cognitive situation a third kind of content, differing from the non-symbolic content since it is symbolic in character, yet not to be identified with the symbolic content since it refers to the relations of applicability holding between the symbolic and the non-symbolic contents in the direction indicated. When we know, we are not merely aware of something, nor are we merely aware of a symbol for something; we are aware also of the symbol for the applicability of that symbol to the thing. Thus in a genuine knowledge situation we are aware of an occurrent, and of a symbol for the occurrent, and of a symbol for the relation between the two.

It must be pointed out that science is the complex inter-relation of these various aspects. Thus, strictly speaking, if any one of the aspects is lacking there is no science. With reference to awareness and content the union seems to be indissoluble, i.e. neither seems to occur apart from the other. But there are cases in which one kind of content is lacking. There are possibly cases in which we are aware of symbols for occurrents in the absence of the occurrents, and there are many cases in which we are aware of occurrents and of symbols for occurrents without being aware of any symbol permitting the attribution of the latter to the former. The question then arises as to whether these are cases of scientific cognition at all. Certainly if they are, they are cases of *inadequate* science, and one may very easily exclude them from the category of science by insisting that inadequate science is not science. This is the safest procedure upon phenomenological grounds; we shall therefore insist that science is not science unless both awareness and content are given, and unless the three types of content are given in the proper inter-relationships.

Thus, in summary, science may be described as an awareness of something which is numerically and qualitatively other than the awareness, and which is related to the awareness in a way which we describe by saying that the awareness is *of* it. This content is a complex consisting of symbolic elements and non-symbolic elements. The symbolic elements are of two kinds: those referring to the non-symbolic elements and those referring to the relations between the other symbolic elements and the non-symbolic elements. If either the awareness or one of the types of content is lacking, there is no science—at least there is no *adequate* science; and if inadequate science is not science, then there is no science.

Science as a Value Activity

But science is more than the mere awareness of a content. Science is a value activity and consequently shares the general features of all such activities. From the point of view of its general place in the cultural life of man, science is the same kind of thing as art, religion, and morality. These activities arise, we have seen, because man is fundamentally a creature of preference who finds a state of affairs which is not to his liking and sets into operation certain activities aimed at the elimination or moderation of these undesirable conditions. Man tries to minimize the negative value and to augment the positive value. When the negative value is of the kind which we call ignorance and error, and the positive value is of the kind which we call knowledge and truth, then the activities which are set into operation are called science.

Thus we may hope to understand something of the character of science as a specific value activity by determining the properties of value situations in general. But many of such considerations lie beyond the scope of the present book. In this class lie all questions pertaining to whether values are objective or subjective, intrinsic or extrinsic, individual or social, whether the activity determines the value or the value determines the activity, what constitutes the differentiating features of the specific values, etc. However, certain more determinate features of value situations may be indicated here, for it is only through the placing of science in its proper valuational context that one can understand certain features of its structure. All value situations seem to exhibit, among other characteristics, the following properties : (1) the recognition of the existence of negative value, (2) the recognition of the existence of positive value, and (3) the recognition of the modifiability of situations exhibiting negative value toward a greater realization of positive value. Let us consider each of these briefly.

(1) The recognition of negative value seems to be a presupposition of the value activity. It is because the individual is impressed with the reality of disease, fatigue, poverty, social conflicts, vice, and ugliness, that he engages in certain activities aimed at their moderation or elimination. The recognition of the absence of a good is a presupposition to the striving for the good. In the value activity which is called science the negative value would be described by such terms as ignorance, error, charlatanism, superstition, mystery, irrationality, ineffability, unsolved and insoluble problems, contradictions, and meaninglessness. All of these are recognized as in some sense undesirable. If science is a value activity, the scientist must be able to point to situations exhibiting this negative value, for it is with such situations that he begins his work.

(2) But the recognition of positive value is also a presupposition of the value activity. The striving after a good which is the essential feature of the value experience would be meaningless unless there were either the imaginative envisagement or the actual discovery in other situations of the good to be attained. Society does exhibit a fair amount of health, recreation, wealth, social harmony, virtue, and beauty. In the value activity which is called science the positive value would be described by such terms as wisdom, truth, justifiable knowledge, verifiable hypotheses, rationality, effability, solved problems, non-contradiction, and meaningfulness. These are more desirable than their negatives. If science is a value activity, it must thus contain indications as to the more desirable form which the activity may take.

(3) Finally, if the situations exhibiting negative value are not such as to yield to modification, then the value activity is futile.

But man finds that disease can be progressively eliminated, poverty lessened, social organization improved, character developed, and ugliness transformed into beauty. He is confronted with the attainment of a good, if not completely at least by approximation. Thus in science man should find that the realization of the positive value is directly dependent upon theories of the knowing process, methods of procedure, instrumental techniques, and principles of logic. By substituting effective methods for ineffective ones, he should succeed in realizing the value more quickly and more surely. Hence science must reveal within its structure such features as will make this progressive attainment of truth possible.

Consequently if we consider science to be a type of value activity, and if we maintain that through an analysis of value activities we may learn something about science, we are bound to conclude that science must exhibit, in some form or other, the three features of value situations which we have just enumerated. Specifically this means (1) that the recognition of the existence of inadequate knowledge must be part of science, (2) that the recognition of the criteria for science's own estimation of itself must be a part of science, and (3) that recognition of the modifiability of science in the direction of greater adequacy must be a part of science. Crudely stated, this means that science must contain within itself its own starting-point, its own goal, and the rules for the attainment of the goal.

But does science actually reveal such data? If our phenomenological analysis is to be accepted, it does not. From this one might draw two alternative conclusions—either that science is not a genuine value activity and does not therefore exhibit the general features of such experience, or that in making our phenomenological analysis of science we excluded such data, *by definition*, in the preliminary demarkation of the field. One suspects the latter to be the case. What one finds and does not find in a situation is determined by the concept in terms of which he identifies and ascertains the scope of the situation. Hence if one defines science so as to limit it to its essential core only, then one should not be surprised to discover that elements which are part only of the larger object are not to be found.

This suspicion may be corroborated if we find that science, in the broader sense of the term, does contain precisely these data.

Clearly, inadequate knowledge is given and is recognizable. It may be of several kinds. Knowledge may be inadequate when either the symbol or the object is lacking from the content of the awareness. In the former case we have a mere awareness of an object without an awareness of its character, an awareness of a *that* without an awareness of a *what*. This may be knowledge, but if so it is ineffable and incommunicable; it has the certainty of the mystic experience but cannot be corroborated or substantiated through cooperative effort. Thus it is inadequate knowledge. In the latter case we have a mere awareness of a symbol without an awareness of any object which could be its referent, an awareness of a *what* without an awareness of a *that*. This may be knowledge, but if so it is incapable of denotative corroboration; it is rather meaningfulness than knowledge, and lacks that feature of applicability which is its essential pragmatic justification. Thus it is inadequate knowledge. Moreover, knowledge may be inadequate when both object and symbol are present, yet the symbol is inapplicable to the object. This is negative knowledge. This, again, may be called knowledge; for to know what a thing is not, is to know something about the thing. But it is relatively inadequate knowledge, for as a rule we wish to know what symbols do apply to an object, not what symbols do not apply. Finally, knowledge may be inadequate because it is unjustified knowledge. In this case there is the awareness of the object, the awareness of the symbol, and the awareness of the symbol for the applicability of the symbol to the object. But there is no further awareness of reasons why the symbol must be applied to the object; thus the certainty which attaches to this knowledge is merely the certainty of individual conviction; the tests of truth have not been recognized or employed. Hence the knowledge is inadequate. Here we are led directly to the second consideration.

The second demand was that science should contain instances of its own positive value. This means that there must be situations in which the knowledge is recognized as relatively adequate.

The most adequate knowledge is that in which there is involved not merely the content of the awareness but also another content which may be called its ground or justification. This content will be symbolic and will be presumed to be applicable to *its* situation. Specifically it will be as follows: " Any symbol which possesses certain properties is applicable" and " The symbol in question possesses these properties." From the awareness of this complex content one may derive the awareness of the content, " The symbol in question is applicable;" hence one justifies the awareness of the applicability of the symbol to its object. Thus there is in many cognitive situations the vague envisagement of the ideal to which all cognition is designed to approximate. If the attainment of truth is possible, then one must know both the marks of truth and the degree to which any given situation possesses these marks. What these marks are need not concern us here, though the question is important for science. It can be readily seen that such tests as " emotional congruity," "immediate conviction," and "support by unrecognized authority " constitute unsatisfactory tests, while " experimental verification," " symbolic consistency," "support by acclaimed authority," etc., constitute satisfactory tests.

The third demand was that science should exhibit the possibilities of its own modification toward greater adequacy. This is clearly the case, since it is true not merely that there are adequate and inadequate systems of science but that inadequate systems tend to yield to more adequate systems. Science is a guided activity in which through the employment of techniques and rules we modify the less adequate symbolic systems and replace them by more adequate ones. Symbolic systems are flexible and may be modeled to fit the world; the ineffable becomes the effable, the incomprehensible becomes comprehensible, and the untested may often be justified. Thus we actually experience the improvement of knowledge; science is not only modifiable but is actually being modified.

It follows that the data which are revealed in the empirical analysis of cognition are somewhat more extensive if one defines science broadly than if one defines it narrowly. The results of this analysis may be conveniently summarized by attempting to fit them into the traditional interpretation of cognition. According to this latter interpretation scientific cognition consists of four elements which may be called respectively the knower, the known, the knowledge, and the knowing. The knower is the scientist, the known is nature, the knowledge is his symbolic system, and the knowing is his method. The general descriptive terminology in terms of the derivatives of the verb " to know " has certain advantages, because it exhibits the elements of cognition in their proper relationship without characterizing them as to further content. It does not demand, for example, that the known should be a world of matter, or of ideas, or of sense qualities, or of appearances ; nor does it insist that the knowing should be intuition, or reason, or experience, or revelation ; nor does it demand that knowledge be psychical, or verbal, or pictorial. Yet it does indicate the essential structural relations of the elements, viz. that knowing is an activity performed by an actor upon a passive subject-matter with a view to the construction of an entity having a desired relation to that subject-matter. This may be expressed crudely by saying that the knower, the known, and the knowing constitute a complex cause which brings about an effect, the knowledge.

According to the point of view suggested in the present chapter these elements may be interpreted as follows : (Since the problem of the nature of the self is irrelevant we shall not attempt to characterize the knower).

The known, or object of knowledge, will be defined in terms of the content of awareness. But it will not be identical with this. That which we *know* is only a part of that of which we are *aware*. We know objects but in order to know them we become aware of symbols. Thus if we are merely aware of objects, or merely aware of symbols, or aware of objects and symbols without being aware of the applicability of the latter to the former, or aware of objects and symbols and of the applicability of the latter to the former without being aware of the justification for this applicability—in all such cases we have what may be called either mere awareness or else knowledge of varying degrees of inadequacy. When the content of awareness becomes sufficiently inclusive to encompass the object, the symbol of the object, the symbol of applicability, and the symbol for the justification, then we have genuine knowledge, i.e. adequate knowledge. Thus what we really know is objects, or, as we shall call them later, occurrents; but we cannot know occurrents adequately unless we employ symbols which are not only applicable but justifiably applicable. Occurrents, we shall see, have structural relations to one another which are themselves occurrents, and in knowing occurrents we thus know the content and structure of existence or nature.

The knowing, or method, will be defined as the process of becoming aware of occurrents, of their symbols, of the symbols of applicability, and of the symbols of justification. Thus knowing is the actual technique for producing in awareness its proper content. This will involve all of the processes of creating situations in which occurrents are given—observational and experimental techniques, instrumental aids, manipulative devices, etc. It will also involve all of the processes concerned with the derivation of meanings through operational routes—naming, describing, generalizing, correlating, etc. Finally it will involve all of the processes associated with the corroboration of the application of symbols to occurrents, i. e., the employment of the tests of truth.

The knowledge, or the result, will be defined as the awareness of a complex content consisting of an occurrent, a symbol, a symbol of the applicability of the symbol to the occurrent, and a symbol of the justification of the applicability. Since unjustified knowledge is usually relatively satisfactory, one need not insist upon the presence of this awareness of an additional complex content which is made up of the test of truth and the fact that the given content exemplifies the properties demanded by this test. Awareness of such a content constitutes justification, though not an absolute justification. For another individual may call into question the tests which I am employing, thus charging me with wrongly applying the tests of truth. This is equivalent to pointing out that I *must justify my criterion for justification*. Clearly this demand for justification may go on indefinitely; actually we rest satisfied if we can agree on the criteria and if we can agree that the symbols in question satisfy these criteria.

The Explanation of Science

Having pointed out the features which are revealed in the empirical analysis of cognition, we might well end our discussion at this point. Science is merely what it is clearly and distinctly revealed to be, and that is all there is to it. But there is reason to believe that we have not yet told the whole story. In many ways science gives hints of features which are not obviously and unmistakably present. If we are to understand the structure of science, we must at least make the attempt to clarify and define these obscurely given entities. For if science is to be explicable, they must be of certain kinds; i.e. if the data of science as listed above are to be derivable from theories, these latter entities must have certain properties.

For our purposes only two of these obscurely given features may be called to attention. If science is to be possible it seems inevitable that (*a*) symbols be such as can be applied and (*b*) objects be such as can be known. We may call these, respectively, the empirical feature of symbols and the rationality of objects.

(*a*) Science may be rendered intelligible if symbols are empirical. It will be the task of the following pages to show precisely what is meant by this. Symbols can be applicable only if their meanings are empirically derived. An empirical theory of meaning is therefore under obligation to show that the operational derivation of meanings from actual situations is such as to guarantee the applicability of the meanings to the situations. The operational theory is based upon a recognition of the flexibility of meanings, and their dependence upon the situation from which they were derived and their manner of derivation. Thus there must be, so far as possible, rules for the derivation of symbols. These rules will characterize the operational techniques through which symbols will be given meanings, and symbols will properly be given only such meaning as is justified by their operational derivation. But under such conditions the determination of the applicability of a symbol is a definite, though not necessarily a simple, process. What this process is can only be revealed in the course of the discussion which is to follow. But such an empirical theory of meaning at least makes reasonable the applicability of symbols to objects. It thus has a distinct advantage over all rationalistic theories, according to which the applicability of our symbols to objects is an unexplainable coincidence.

(b) Science may be rendered intelligible if objects are rational. We can know objects only if they are such as can be known. Objects are independent of us but choose to manifest themselves in ways congenial to our rationality. The rationality of objects is not a demand of our understanding, but a discovered feature of the world. We should not know what rationality meant unless there were hints of it in the given. Objects exhibit certain features which seem to favor the act of symbolism. These are three in number: (1) independence of will; (2) publicness; (3) fixity.[1]

(1) Independence of will is the property of objects by virtue of which we say that they have been discovered. Acts of will are given just as other objects; when acts of will are given along with objects, we tend to say that the objects have not been discovered but have been created. When objects impress themselves suddenly and forcibly, without apparent antecedents in previous awareness, we say that they exist independently of the will. (2) Publicness is the property of objects by virtue of which communciation is possible; denotative gestures have significance because nature is common to minds. Some occurrents are such as to elude denotative gestures; we then say that the occurrents are more private than public. It is an empirical fact that we do talk successfully about objects; this capacity to be talked about by different individuals is the publicness of objects. (3) Fixity is the property of objects by virtue of which they persist for attention. Some objects possess a relatively high degree of fixity, others possess a low degree. It seems possible to "hold certain objects in mind" and to recall objects. This is what we usually call the uniformity or fixity of objects. The important thing to remember is that this is merely an empirical fixity, i.e. an observed persistence and recurrence.

[1] Cf. Norman Campbell, *What is Science* (London, 1921), pp. 20–36; also *Physics, the Elements* (Cambridge, 1920), chap. I; A. D. Ritchie, *Scientific Method* (London, 1923), pp. 23–30; F. Barry, *Scientific Habit of Thought* (New York, 1927), chap. II; H. Dingle, *Science and Human Experience* (London, 1931), pp. 17–20, also chap. IX; R. Carnap, *Der logische Aufbau der Welt* (Berlin, 1928), pp. 90–1.

But this does not seem to constitute an adequate definition of rationality, for it defines rationality "intra-experientially," i.e. it attempts to determine the character of objects *in themselves* by an examination of their character *in the context of cognition*, hence it does not adequately meet the problem which it was designed to solve. Science is possible only if objects exhibit in themselves the same rationality which they exhibit in the context of cognition. But when the problem is defined in this way, a strictly empirical solution becomes impossible, for we are inferring from the character of an object in experience its character outside of experience—and the principle of this inference cannot itself be empirically justified. Hence when we define rationality in these terms, we must characterize it as the hypothesis of rationality rather than as the construct of rationality.[1] This gives us the features of absolute objectivity, absolute publicness, and absolute constancy. The hypothesis of absolute objectivity is that objects which occur in association with awareness may be identical with objects which occur in dissociation from awareness. Thus the same object may or may not enter into awareness, and the fact of its entering into awareness does not change its character; awareness makes no difference to an object and is external to it. The hypothesis of absolute publicness is that objects which occur in association with one awareness may be identical with objects which occur in association with other awarenesses. Thus the same object may enter into the awareness of different individuals. The hypothesis of absolute constancy is that objects which are associated with an earlier awareness may be identical with objects associated with a later awareness. Thus the same object may recur for awareness.

This gives the external, public, and uniform world which is presumed by all science. Such a world cannot apparently be derived from the intra-cognitive world, but it may be considered as a plausible hypothesis in terms of which science may be understood. It is my own view that there are many aspects of the hypothesis of rationality which can be reduced to the construct of rationality, though the problem is too extensive to be

[1] This distinction will be explained in chapter IX.

discussed adequately at this point,[1] and the topic is not vital to the question of the logical structure of science.

I have attempted in this chapter to give a sketch of the general structure of the scientific cognitive situation. We are now ready to turn to a more detailed examination of certain features of that structure.

---

[1] Cf. W. T. Stace, *Theory of Knowledge and Existence* (Oxford, 1932) esp. chap. VI ; also Carnap, *op. cit.*

CHAPTER III
NATURE: OCCURRENTS

It hardly seems necessary to offer an apology for the introduction of metaphysics into a consideration of the logical structure of science. Science is avowedly empirical in its aim and intent, and its task is that of devising a system of symbols representative of the structure and content of the realm of nature. Hence the logical character of its doctrines will be a reflection of the metaphysical character of its subject-matter. The brief consideration of the structure of nature which follows is, therefore, a necessary preliminary to the consideration of the specifically logical problem.

Since the problems here treated are only incidental to the development of my general thesis, I shall take the liberty of being both dogmatic and brief. I shall not in most cases give reasons for my statements, nor shall I take the space for the extensive elaboration and illustration which are helpful in the elucidation of this type of abstract principle. I feel that such procedure will not lead to confusion and obscurity, because I believe that the metaphysical principles to be formulated are essentially those which lie at the basis of the common sense view of the world. Science herself proceeds upon the assumption of a common sense outlook and modifies it only when inconsistencies are revealed. I have tried, therefore, to formulate this view of nature with some of the most glaring of the inconsistencies removed, and with such further transformations as are demanded by the conclusions of modern science.

Occurrents

We have already seen something of the features of nature. In the last chapter we indicated the reasons for supposing that nature possesses certain formal properties of objectivity, publicness, and constancy. But these properties are not very highly determinate as characterizations of nature; conceivably there might be many natures which manifest these formal properties and yet which would differ fundamentally from ours. Hence we must endeavor to point out the more specific properites of nature. When we look about for a term by which to designate the entity which may be said to be the basic subject-matter of science, we find many candidates : event,[1] continuant,[2] object,[3] substance,[4] sense-datum,[5] particular.[6] The term which I have chosen is " occurrent."[7] Occurrents make up the stuff of nature. Nature is a complex of occurrents and is, presumably, also itself an occurrent. The notion of occurrent is one of the basic indefinables of philosophy. In order to define it one would be obliged both to find a term of wider denotation and to specify the feature differentiating it from its coordinate species. One might say that occurrents are entities ; but this conveys no important information for the term " entity " is essentially vague, and any attempt to make it more precise would naturally result in a gesture of pointing to such a thing as an occurrent. Perhaps the most satisfactory definition of the term is one which asserts that an occurrent is anything that can be said to be or to be happening in the ordinary senses of these terms.

An example of an occurrent may be found by taking any spatio-temporal increment of reality. This is not a way of defining an occurrent, for space and time are themselves occurrents ; but it is a convenient means for locating an occurrent. Examples of occurrents will be such as the following : The table on which I am writing for any specified time—say a certain five minutes during a given day—, or the drawer of the table for such a time or for a longer or shorter included or including time, the total events of my life for a given day, the life of a leaf on the tree outside of my window for a given three months,

---

[1] Whitehead, *Concept of Nature* (Cambridge, 1920), p. 19 ; Russell, *Philosophy* (New York, 1927), chap. XXVI ; also *Analysis of Matter* (New York, 1927), p. 286.
[2] W. E. Johnson, *Logic* (Cambridge, 1921), Vol. I, p. 199.
[3] L. Wittgenstein, *Tractatus Logico-Philosophicus* (London, 1922), prop. 2.02 *et seq*. Cf. also Whitehead, *op. cit.*, chap. VII.
[4] Wittgenstein, *op. cit.*, prop. 2.024 ; B. Bavink, *The Natural Sciences* (New York, 1932), p. 2.
[5] Russell, *Our Knowledge of the External World* (Chicago, 1915), lect. III ; Broad, *Scientific Thought*, p. 240.
[6] Russell, *Mysticism and Logic* (London, 1921), p. 129 ; *Analysis of Matter*, p. 277.
[7] I have taken the term, though not its meaning, from Johnson, *op. cit.*, Vol. I, p. 199.

a river for a given period of 1000 years, the World War, the centaur which I am now thinking about, my toothache for a ten minute period, etc. All that is required is to specify a certain block of space and time and designate what is happening in that volume. One need not know what is happening; thus one can specify an occurrent simply as "What is happening here-now." Occurrents may be inclusive or limited spatially or temporally; hence occurrents may include and be included in other occurrents in both spatial and temporal ways. Or diverse occurrents may happen in the same block of space and time. Or diverse occurrents may partly overlap in spatial or temporal ways.

It is convenient to distinguish the *form* of an occurrent from its *content*. We may say that the form of an occurrent is that by virtue of which it is an occurrent. This does not mean that there is an abstract form existing apart from the occurrent which materalizes itself into the concrete occurrent. It means rather that all occurrents have a general character, and that one can grasp immediately that he is confronted with an occurrent by recognizing this character.[1] But every occurrent has a certain individuality which distinguishes it from every other occurrent; this may be called the content of the occurrent. Thus an occurrent is *this* occurrent rather than *that* occurrent, but it exemplifies the general form of occurrents and thus shares something with all other occurrents. But since occurrents are indefinable, it is impossible to say what this common feature is. Awareness of an occurrent tells us that it is an occurrent of which we are aware, i.e. that it is a particular occurrent and has the general form of all occurrents. However, we may be abstractly aware of the form of an occurrent without being aware of its content. Such awareness as this is involved in any discussion, such as the present one, which attempts to portray the general character of all occurrents. We must be able to see not only the specific features of this occurrent and that occurrent; we must become aware of the abstract form of all occurrents distributively (and collectively also, if the totality of occurrents is an occurrent).

It is convenient also to distinguish between the individual or *unique content* and the *structural content* of an occurrent. Every

[1] Cf. Wittgenstein, *op. cit.*, props. 2.024 and 2.025.

occurrent has relations to other occurrents. Such relations constitute a part of the content of the occurrent, for to state any of its relations is to state something about it. In this sense all relations are internal, and to characterize completely the structural content of an occurrent is to refer to every other occurrent. But in order to talk about any occurrent it is not necessary to mention all of its structural relations; thus it is possible to make intelligible propositions about parts of reality. By the unique content of an occurrent is meant that which possesses the structural content. Wherever there is a relation there will be at least two unique contents. All relations of the occurrent are external to the unique content but internal to the structural content. One may become aware of the unique content without being aware of the structural content; thus in order to know an occurrent it is not necessary to know everything else in existence. The unique content cannot be characterized in the ordinary way, i.e. by pointing out relations of difference and similarity to other occurrents; for this would involve the introduction of structural relations and these are not part of the unique content. But the content of an occurrent can be characterized in that it can be symbolized by a meaning having a correspondingly unique content; hence the unique feature of the meaning reveals the unique feature of the occurrent, and the structural features of the meaning reveal the structural features of the occurrent.

Relational Occurrents

Relations between occurrents are also occurrents, though they are occurrents of an essentially different kind. However, the difference between relational and non-relational occurrents is inexpressible except by question-begging terms. One may say that a relational occurrent is one which connects, or ties, or goes between occurrents of a non-relational character. This is clearly a circular definition. Or one can show that relational occurrents are those which require non-relational occurrents for their existence, but one cannot then go on to show how they do. It is clear that the terms related do not depend for their existence upon the relations in the same way that the relations depend on the terms. Putting it crudely, we may say that terms are always essential in the consideration of relations, but relations are not always essential in the consideraton of terms. A relation must always connect terms, but terms need not be related essentially.

The existence of relational occurrents is correlative with the fact that there are complex occurrents. For wherever there is a relational occurrent, there is thereby designated a more inclusive occurrent of which the terms and the relation are parts. This is true of all relations, not merely those which designate spatial and temporal connections but those of similarity, difference, degree, etc. But the latter type are unimportant from this point of view, because the whole which is designated does not have important relations to other occurrents; i.e. it does not act as a single occurrent. To put it in other words, every relation between occurrents reveals the structure of the realm of occurrents, but some of these relations are more basic than others, just as some properties (space, time, shape, size, weight, number, etc.) are more basic than others. Hence it is not the task of metaphysics to enumerate all such relations; it suffices to mention and describe the most basic, just as it suffices for the metaphysician to mention only the most basic properties of things.

It follows that there are degrees of complexity, kinds of complexity, etc. in reality. The universe as a whole must be an occurrent; this will be the only occurrent which does not have relations to occurrents outside itself in a spatial or temporal sense, though it will have such relations to occurrents inside itself. And there will be wholes which are spatially complex, wholes which are temporally complex, and wholes which are complex in neither sense. A whole which is spatially complex with reference to certain elements may be non-spatially complex with reference to other elements. And so on.

The most basic and general of all relations between occurrents is that of *otherness* or *difference*. Such a relation as this must exist if we are to speak of occurrents (in the plural), if we are to speak of relational occurrents, and if we are to speak of complex occurrents. In what ways, therefore, may occurrents differ from one another?

They may differ, first, in unique content. This is what is ordinarily called qualitative difference, provided the word is used so as to include quantitative (though not mere numerical) difference. It refers to any difference in content which is not reducible to spatial or temporal relations of the occurrents in question, since qualitative diversity may be found even when the occurrents are spatially and temporally identical. Thus qualitative difference is the relation which a shape bears to a color, or a taste to a sound, or a number to an odor, whether or not the occurrents are found united in a single spatio-temporal situation. But two occurrents may differ, secondly, by virtue of their respective locations; in this sense they differ from one another not through unique content but through structural relations of a certain kind. Structural difference may be found associated either with qualitative similarity or with qualitative difference. Thus a blue which is here differs from a blue which is there, and a sound which is now differs from a sound which is then; but a blue which is here or now also differs from a sound which is there or then. Thus two occurrents may differ as to unique content and not as to structural relations, or they may differ as to structural relations and be either alike or different as to unique content. It is clearly impossible for two occurrents to be qualitatively alike and also structurally alike, for they would then be identical and could not be spoken of as two.

Association and Dissociation

Relations of otherness determine the basic complexity of the realm of occurrents. But relations of otherness occur in two general forms, viz. relations of *association* and relations of *dissociation*. The relation of association may be defined loosely as the togetherness of occurrents in a space and time. More accurately, it is the relation of a certain occurrent to a spatial occurrent which is its space, or of a certain occurrent to a temporal occurrent which is its time, or, derivatively, of certain occurrents to each other by virtue of common associative relations to a spatio-temporal occurrent. For example, the occurrent which is this table is associated with the space which it would ordinarily be said to occupy and with the time during which it would ordinarily be said to occur. When we say that an occurrent is *at a space* or *at a time* we are using an elliptical expression for the statement: the occurrent is associated with another occurrent which is a space-time (or with another occurrent which is a space and another occurrent which is a time). Derivatively two occurrents are associated together when they are associated with the same space-time; for example the white of the sugar is associated with the sweet, and the smooth of the table is associated with the hard.

When we assert that an occurrent which is not a space is associated with an occurrent which is a space, we allow for the fact that the non-spatial occurrent may be a time. Thus an occurrent which is a time may be associated with an occurrent which is a space, and conversely. This associative complex of a space and a time is very important in determining the location of an occurrent. Let us call such an associative union of a space and a time, a *situation*. Then a situation is a more or less extensive spatio-temporal block of the realm of occurrents. It may be extended spatially and limited temporally, as in the case of the universe for a second; or it may be limited spatially and extended temporally, as in the case of a cubic centimeter of gold for a century. The problem of determining what space and time an occurrent occupies will be expressed in our terminology as the problem of determining with what situation an occurrent is associated. Since it is an empirical fact that nothing ever happens at a time without also happening at a space (although the spatial associate may be precise and the temporal associate vague, and conversely), it follows that we always have given not merely a space or merely a time, but a situation. We may then say that the situation with which an occurrent is associated is the situation *of* the occurrent. It seems possible to specify a situation without specifying the qualitative occurrent associated with the situation, though it may be necessary to indicate the qualitative occurrents which are boundaries of the situation. What is more important, it also seems possible to specify the absence from a situation of a kind of occurrent known in advance; thus it is possible to know, for example, that a given situation is not red without knowing that it is some color other than red. It follows that the notion of situation is important not only in locating occurrents but also in specifying their absence.

The other basic relation between occurrents is that of *dissociation*. This is the relation of any two different spatial occurrents to one another, or of any two different temporal occurrents to one another. Thus the " here " is dissociated from the " there," and the " now " is dissociated from the " then." Two situations will be dissociated if they differ in space or time or both ; thus the " here-now " is dissociatied from the " here-then," and the " here-now " is dissociated from the " there-now," and the " here-now " is dissociated from the " there-then." Derivatively, dissociation is the relation of any occurrent to a spatial or temporal occurrent which is not its own space or time, and of any two occurrents to one another when they are respectively associated with dissociated situations. Thus if a red is here-now and a blue is there-now, the red and blue are dissociated ; if a sweet taste is here-now and a bitter taste is here-then, the sweet and bitter are dissociated ; finally, if a red is here-now and a bitter taste is there-then, the red and the bitter are dissociated. It is clear that the words " association " and " dissociation " are given technical meanings in this connection, though the meanings are not fundamentally different from those found in ordinary discourse. We commonly speak of associations of coexistence, and this is essentially the meaning here intended if one means to specify co-spatiality and co-temporality, i.e. identity of situation. What are usually called associations of sequence, will here be called dissociations, though this need not imply any lack of dependence or inter-dependence with reference to the elements.

We may now see what is meant by saying that an occurrent has the relation of otherness to a different occurrent ; i.e. we may now see the three significant ways in which occurrents may differ from one another. Any two occurrents which are distinguishable are others of one another. But they may differ as to situation (space, or time, or space and time) or they may be associated with the same situation but differ as to unique content.

Thus they may be qualitatively alike but dissociated, qualitatively different and either associated or dissociated. Clearly, they cannot be qualitatively alike and associated, for they would then be identical.

The general features of these basic relations of association and dissociation may now be discussed. (a) Symmetry. Both relations are symmetrical; i.e. it makes no difference whether we say that a red is situated or that a situation is red. So also it is of no concern whether we say that a red is triangular or that a triangle is red; the fact that the latter expression is favored linguistically is due to our use of the term "triangle" to designate the complex occurrent of which "red" is a part; i.e. we use "triangle" substantively and "red" attributively, but this does not prevent each of the terms from having a substantival use. In the case of dissociation, linguistic complications obscure the symmetry of the relation; we generalize the occurrent expressed in the predicate of our proposition and lose the specificity of its reference. For example, to say that a red is dissociated from a certain situation is equivalent to saying that this certain situation is dissociated from a certain (i.e. the same) red. But to say that the dissociation of a red from a certain situation is equivalent to the dissociation of *any* red from this situation is false. The difficulty can be better illustrated in terms of two qualitative occurrents which are dissociated, e.g. a blue and a circle. To say that a blue is dissociated from a circle is not the same as to say that the blue is not circular, or that the circle is not blue. For a blue may be dissociated from a certain circle and yet be associated with another circle, and a circle may be dissociated from a certain blue and yet associated with another blue. It is this important fact which we shall consider later[1] as a ground for deciding that for the logic of science the relations of participation and non-participation of a symbol in a situation are more convenient than the strictly occurrential relations of association and dissociation. All that must be remembered in this connection is that the relation of dissociation is symmetrical if one does not generalize either of the occurrents so connected.

[1] Chapter VII, p. 156.

(b) Transitivity. If we exclude relational occurrents, association is transitive but dissociaton is not. If a red is associated with a situation and this situation is associated with a certain triangle, then the red is associated with the triangle. Thus all of the occurrents associated with a given situation may be spoken of as co-associates. But if a certain blue is dissociated from a certain situation and this situation is dissociated from a certain circle, we cannot conclude that the blue is dissociated from the circle; such may be the case, but it is not without exception. Hence dissociation is non-transitive. The relative products of association and dissociation may be simply determined. If an occurrent is associated with another occurrent which is dissociated from a third, then the first and the third are dissociated. Thus two or more associations determine an association; one or more associations with a dissociation determines a dissociation; and two or more dissociations determine either an association or a dissociation.

(c) Reflexiveness. If any relation at all may be reflexive,[1] and it is therefore possible to speak significantly of the distinction between reflexive and irreflexive relations, we shall be obliged to say that both associations and dissociation are irreflexive. For both relations "imply diversity;"[2] i.e. association implies qualitative diversity, and dissociation implies either spatial or temporal diversity. Thus if any two occurrents are either associated or dissociated, they must be in some sense distinguishable; hence no occurrent could be either associated with or dissociated from itself.

(d) Connectivity. The disjunction "association or dissociation" is connected[3] with reference to the realm of occurrents. This means that given any two occurrents they will be either associated or dissociated. Occurrents which bear relations of inclusion, or of partial overlapping in either space or time will be dissociated. Thus one may say that any two occurrents either do or do not occupy the same situation.

This suggests that space and time are very important concepts

---

[1] See the interesting discussion of this topic by D. H. Parker, *Philosophical Review*, Vol. XLII, pp. 303 *et seq*.
[2] As this notion is defined by Russell. *Vide Introduction to Mathematical Philosophy* (London, 1920), p. 32.
[3] Cf. Russell, *ibid*.

in the determination of the structure of nature. Hence we shall be obliged to present, briefly, the general point of view which is taken with reference to these notions.

Space and Time

Space and time are properties of occurrents (in the ordinary sense of this expression) just as colors, sounds, tastes, shapes, and weights are. This means that space and time are *kinds* of occurrents just as these other qualities; i.e. there are spatial and temporal occurrents just as there are occurrents which are colors, sounds, etc. We discover occurrents which are colors, sounds, tastes, spaces, and times. Spatial occurrents are instances of space in general just as colors are instances of color in general; temporal occurrents are instances of time in general just as sounds are instances of sound in general. Space and time are not forms of apprehension or of perception; they are givens to be analyzed and correlated with other givens. The world is given as spatial and temporal just as it is given in any other qualitative manifestation.

The tendency to consider space and time as unique features of the world is based upon their privileged position. This can be seen to reside in two features. In the first place, space and time are, so far as we can tell, omnipresent;[1] we never observe an occurrent which is not associated with a space and a time. Thus space and time are convenient instruments for locating occurrents (principles of individuation, they are often called in philosophical terminology). Also by virtue of associations of occurrents we may ascertain what kinds of occurrents are occasionally found together in a situation, what kinds are always found together, what kinds are occasionally found dissociated, and what kinds are always found dissociated. These are very important bits of information, for they determine the *laws* of the behavior of occurrents. It is only by knowing with what occurrents a given occurrent is associated, and with what

[1] If quantum phenomena should entitle us to conclude that space-time is rather the frame of our sense-perceptions than of nature herself (*vide* Sir James Jeans, *The New Background of Science* (New York, 1933), pp. 259–60) the omnipresence of space and time would still be a fact. For what nature is in herself must be an inference from what nature is in sense-perception; hence the starting point of science must be a spatio-temporal world.

occurrents it is dissociated, that we can determine its structural content; thus so far as our knowledge of the world is a knowledge of its organization, we must employ spatial and temporal reference.

In the second place, space and time are important features of the world because each of them is internally a structural complex. The instances of space have spatial relations with one another, and the instances of time have temporal relations with one another. These are the ordinary relations of extension, distance, and direction. By virtue of these relations occurrents associated with the respective spaces and times take on derivative relations. Thus the " here " of my typewriter has relations of distance and direction to the " there " of the door; and, as a result, my typewriter takes on these relations to the door. Again the " here " of my typewriter includes the " here " of one of its keys; consequently the typewriter includes the key. So also the " now " with which my writing is associated is twenty-four hours later than the " then " with which my writing yesterday was associated; consequently my writing today is twenty-four hours later than my writing yesterday. It can be seen, therefore, that occurrents which are not spaces and times take over the spatial and temporal relations of the situations with which they are associated, and we have another source for our information of the structural relations of occurrents.

This suffices for a discussion of the *elements* of nature. We now turn to a consideraton of *complexes* of occurrents.

CHAPTER IV
NATURE: COMPLEXES

Nature is the totality of occurrents, but it is an organized totality. Hence, in the consideration of nature it does not suffice merely to mention the elements; one must also show how the elements are structurally united into complexes. But complexes are impossible unless there are elements which are themselves structural in character. These elements we found to be relational occurrents. Relational occurrents connect non-relational occurrents into complexes. The most important of the relational occurrents are associations and dissociatons, and the most significant of dissociative relations are those which spaces bear to other spaces and times to other times. By virtue of the fact that space and time are omnipresent and reveal significant structural properties the entire realm of occurrents takes on structural organization.

Associative Complexes

One of the most important structural features of the realm of occurrents is the fact that the association of two or more occurrents determines another occurrent which is precisely the associated complex. Hence, we may say that certain occurrents are complex in the sense that they " contain " sub-occurrents.[1] If a red is associated with a triangle, then there is formed a complex occurrent, the red-triangle, which is nothing more than the associative union of the two. Most of the ordinary occurrents of life are complexes of this character; we call them *objects* or *things*. A table, for example, is the associative union of a shape, a color, a weight, a smooth, a hard .... and a situation; sugar is the associative union of a sweet, a white, a powder, a brittle, a weight .... and a situation.

The first remark to be made about such complexes is that they constitute what are usually called *concrete* as opposed to *abstract*

[1] The sub-occurrents are then obtainable from the complex occurrent by what H. W. B. Joseph calls " metaphysical division." Cf. *Introduction to Logic* (Oxford, 1916), p. 132; also R. M. Eaton, *General Logic* (New York, 1931), p. 292; the term here used is " qualitative analysis."

occurrents,—at least according to one of the many meanings of this dichotomy. In this sense concrete or complex occurrents " contain " abstract or simple occurrents. (It must be noted that this sense of " containing " is not to be confused with the sense in which an extended time includes a time, or an extended space includes a space, or the occurrent associated with an extended space or time includes the occurrent associated with the included space or time. This is a specific type of inclusion, though possibly reducible to the type here considered. Thus we shall not say that a table is concrete with reference to its leg, nor that the total life of a man is concrete with reference to his youth. The difference between the two types of inclusion can be seen in the fact that a concrete occurrent and its included abstract occurrent are associated with the same situation, whereas an extended spatial or temporal occurrent and its included occurrents are associated with different situations and are therefore dissociated from one another.) Since the abstractness or concreteness of an occurrent is determined by its relation to another occurrent, the distinction between the two is relative; thus an occurrent will be concrete with reference to its included occurrents which are abstract ; but these, in turn, will be concrete with reference to their included occurrents. However, it is possible to point to relatively highly complex occurrents and to relatively highly abstract occurrents. A highly complex occurrent will be one which contains a large number of diverse sub-occurrents and is not itself included in any more complex occurrent; a highly simple occurrent will be one which is contained in a large number of diverse occurrents but does not itself contain any sub-occurrents. The most concrete occurrent possible will be the universe, not in the sense of the most extended spatial and temporal whole, but in the sense that the universe is the ultimate subject of all predications and cannot itself be predicated of anything. Highly abstract occurrents are not easy to find, for we continually discover that occurrents thought to be ultimate simples are really complex ; e.g. a tone may be thought to be simple until one has had attention called to the distinction between its loudness, its timbre, and its pitch, whereupon the occurrent takes on a corresponding complexity and its included occurrents become simples. But it is likely that dimensionality is even more abstract than any of these; hence this becomes the ultimate simple. Although we cannot easily specify any ultimate simples, we can assert that such basic concepts of metaphysics as occurrent, quality, relation, time, and space, designate occurrents possessing a high degree of abstraction. In ordinary parlance we should say that these occurents characterize many occurrents but cannot themselves be characterized by many occurrents. One of the commonest sorts of complex occurrents is the association of a qualitative occurrent with a situation; here the qualitative occurrent is abstract with reference to the associative union of the qualitative occurrent and the situation, which is concrete. This explains why such terms as "thing" and "object" designate concretes rather than abstracts; whenever we use such terms as these we tend to think of a qualitative occurrent in a situation, whereas whenever we think of the attributes of things we tend to think of the sub-occurrents in isolation from the situation.

The second remark about associative complexes is that they are not readily identified and are commonly mistaken for some of their abstractions. So far as I can see, much of the debate about synthetic and analytic propositions centers about this fact. If a concrete occurrent is the togetherness of its sub-occurrents, then there is no way of recognizing the concrete occurrent apart from these elements. And if the occurrent is highly complex, i.e. it is the associative union of a large number of occurrents, the occurrent as a whole can be identified only when each of the elements has been located. The result is that we tend to identify the whole rather with one or more of its elements; these elements are then called the essentials of the whole, and any statement that the whole contains these parts is analytic and necessarily true. Thus it is an analytic proposition that gold melts at $1075°$, since this is taken as the identifying property of gold. But as more sub-occurrents are discovered in the associative complex, the term designating it takes on new content. Each affirmation that the whole contains one of these newly discovered parts is a synthetic proposition. But as it becomes recognized that the newly discovered associate is intimately connected with its associates, it takes on the character of an essential element; henceforth the assertion that the whole contains this element is an analytic proposition. Thus if it is discovered that the melting point of 1075° is associated with a specific gravity of 19.3, then the proposition that gold has a specific gravity of 19.3 becomes an analytic proposition. Expressing the same thing symbolically one may say that if an associative complex represented as (a-b-c-d) is designated as K, then one may select a as the mark of the complex and define K as that which exhibits a. Then if it is later discovered that the complex is more accurately represented as (a-b-c-d-e), the assertion that K possesses the attribute e is a synthetic proposition. But then K is seen to be constituted by the larger complex; hence to assert that K possesses the attribute e gradually becomes an analytic proposition. Whether or not a proposition is analytic is dependent upon the fluctuating scope of the subject term; and this, in turn, is due to the difficulty in grasping complex occurrents. Highly abstract occurrents exhibit the difficulty to a lesser degree, since their complexity can be quite completely grasped; hence propositions about abstract occurrents tend very readily to become analytic.

But, in the third place, it should be pointed out that there is a corresponding elusiveness about high abstractions. High concretions do not enter readily into awareness, and, by contrast, their elements are easier to grasp; for example, a patch of red, a sweet taste, a bitter odor, a sensation of heat, seem to be the most indubitable elements of experience; one feels that he can never be genuinely mistaken about them. But certain of these simples have only an apparent simplicity, and reveal by analysis higher abstractions of which they are composed. These abstractions of abstractions seem to be even more elusive; one cannot be sure that he is experiencing them nor what precisely he is experiencing. It is not so easy to become aware of the pitch, timbre, and loudness of a tone as it is to recognize the tone itself as a complex; it is harder to grasp the asymmetry of a relation than to detect the relation as a more concrete entity. The detection of abstract occurrents seems to require a certain type of mind or training.

The fact that high concretions and high abstractions are not readily detected in the realm of occurrents is a very important fact for science. It means that science must start and end with "middle-sized" occurrents, and that all reference to high concretions and high abstractions must be through hypotheses and constructs. If occurrents are given obscurely, then we must determine their contents through the situations in which they are found,—viz., through the less obscure (less highly concrete and less highly abstract) occurrents with which they are associated. This means that the properties of highly concrete occurrents must be inferred from the properites of their elements, and that the properties of highly abstract occurrents must be inferred from the properties of the complexes in which they are found. But the elements of the highly concrete occurrents are precisely the complexes of highly abstract occurrents; these are the " middle-sized " occurrents which constitute the ultimate data for science. It is one of the problems of philosophy to determine what these ultimate data are, whether they are what would commonly be called relatively high abstractions, such as number, space, and time, or what would commonly be called relatively high concretions, such as organism and mind, or what would commonly be located intermediary between high abstractions and high concretions, such as matter and sense-data.[1]

Dissociative Complexes

Another structural feature of the realm of occurrents, as important as the fact of associative complexes, is the fact of dissociative complexes. A dissociative complex is a complex which is formed by the dissociative union of two or more occurrents. Thus the essential feature of a dissociative complex is that it is made up of occurrents which are spaces or times or of occurrents which are associated with different spaces or times; a dissociative complex might consist of situations or it might consist of occurrents having diverse situations. The most important of the dissociative complexes are those in which there is an identity of space with diversity of time, as in the history of a mountain, or an identity of time with a diversity of space, as in a society of human beings at a period.

[1] Cf. Carnap, *Unity of Science* (London, 1934), pp. 45–7.

The structure of dissociative complexes exhibits many features analagous to those of associative complexes. We may therefore follow our treatment of these complexes in our remarks about dissociative complexes.

In the first place, analagous to the distinction between concrete and abstract associative complexes, there is the distinction between extensive and minute dissociative complexes. Extensive occurrents " contain " minute occurrents. This is the sense of " contain " which is applicable to all whole-part relationships, either of space or of time, and is identifiable with Whitehead's use of " extension."[1] Thus we may say that a dissociative complex extends over any other dissociative complex which it contains either as a spatial or as a temporal element. For example, the dissociative occurrent which is a table for a minute contains the occurrent which is the leg of the table for that minute, and this in turn will be a dissociative complex with reference to any part of the leg for that minute. Moreover, the dissociative occurrent which is the table for a minute contains the occurrent which is the table for any part of that minute, and this in turn will be a dissociative complex with reference to the table for any part of that portion of a minute. A highly extensive dissociative complex will be one which contains many spatial and temporal parts and is not itself contained in any more extensive whole. The most extensive occurrent is the universe, as that which contains everything. Very minute occurrents usually yield to spatial or temporal analysis ; i.e. complexity is discovered through the use of scientific instruments permitting the observation of the more and more limited spatially and the more and more brief temporally.

In the second place, highly extensive dissociative complexes are not readily observable and are commonly mistaken for their partial occurrents. This fact seems to me to throw much light on the controversy as to the validity of analysis and the possibility of " reducing " a whole to its elements. A whole is always the complex of its elements, and to say so is to utter a strict tautology. Correspondingly, to say that a whole is " more than "

[1] *Concept of Nature*, p. 58 ; also chap. IV ; Cf. Carnap, *Der logische Aufbau der Welt*, p. 48.

the complex of its elements is to utter a contradiction, although a whole may be more than the "sum" of its elements, or more than any one of its elements. The point of this dispute seems quite clear when one recognizes that certain wholes are so extensive that one may become aware of the elements much more readily than he can become aware of the whole; in all such cases there has been progress in the direction of understanding when the whole is "reduced" to its elements. The only way to understand society is to understand human beings in society; the only way to understand the operation of a machine is to analyse it into its elemental movements; the only way to understand the universe as a whole is to approach it in a piece-meal fashion. But in adopting this procedure one should recognize that there are elements of structure which are almost certain to be lost in the synthetic operation which produces the whole from the parts. In this sense the whole is probably more than the mere juxtaposition of the parts, and to identify the two is to lose a feature of the whole. Thus the fact that extensive wholes are elusive and better understood in terms of their elements should not lead to the belief that the wholes are merely the elements in their *known* relationships.

But in the third place, there is a corresponding elusiveness about minute occurrents. Thus we are obliged to supplement the preceding principle by another: Highly minute occurrents are not readily observable and are commonly mistaken for the more extensive occurrents of which they are parts. This again bears upon the validity of analysis. But here the analysis seems to be highly artificial, because the whole is clearly given and immediately understood, while the elements of the whole are obscure and hypothetical; thus to explain the whole in terms of the properties of its elements seems to be a process of explaining something which we already know in terms of something which we don't know. Doesn't this account for the feeling of artificiality which one often experiences in the presence of an explanation of colors in terms of wave lengths, heat in terms of molecular motion, or matter in terms of space-time deformations? And doesn't it also account for the tendency (against which the scientist has to be continually on his guard) to attribute *color* to the *waves*, *temperature* to the *molecules*, and *shape* to the *space-time deformation*? Here the danger is not in attributing to the whole the properties of the parts, but in attributing to the parts the properties of the whole.

Thus the question of dissociative wholes is also very important in the determination of the ultimate data of science. If the highly extensive occurrents and the highly minute occurrents are not readily detected in the realm of occurrents, science must clearly start and end with " middle-sized " occurrents, and all reference to the extensive and the minute must be through hypotheses and constructs. Whenever occurrents are given obscurely, we must determine their contents through their contexts. This means that the properties of highly extensive wholes must be inferred from the properties of their elements, and that the properties of highly minute occurrents must be inferred from the properties of the dissociative complexes of which they are parts. Thus the moderately extended occurrents constitute the explanatory stuff both for the more and for the less extended occurrents.

But, as we saw, science is also concerned with " middle-sized " occurrents in another sense, viz., those intermediary between the highly concrete and the highly abstract. Thus so far as complexes are concerned we may say that the data of science are means along the dimension concrete-abstract, and along the dimension extensive-minute. Complexes located at the extremes of either of these dimensions are to be defined in terms of the means, since it is only the latter which are clearly given. We shall return to this topic immediately. But before leaving the subject of complexes it is necessary to say something about one particularly important type of complex,—viz., a *class*.

Complexes may vary along a dimension which may be called " homogeneous-heterogeneous." This dimensional variation is determined by the degree of similarity exhibited by the elements of the complex with reference to one another. A complex is essentially homogeneous when its elements are significantly like one another, and essentially heterogeneous when its elements are significantly unlike one another. The similarity and dissimilarity may be spatial or temporal, or they may be qualitative. This results in a great variety of kinds of complexes. If a complex occurrent exhibits spatial homogeneity but temporal heterogeneity, then it is a historical object at a place, and it is a changing or a permanent historical object depending upon whether it exhibits qualitative heterogeneity or qualitative homogeneity. If a complex occurrent exhibits temporal homgeneity but spatial heterogeneity, then it is a non-historical object which is extended in space, and it is heterogeneous or homogeneous depending upon its qualitative similarity or diversity. If a complex occurrent exhibits both spatial and temporal homogeneity, then it is an associative complex and cannot be perfectly homogeneous in a qualitative way since it would then cease to be complex; however, it might be relatively homogeneous, as in the case of a musical chord, where there is a temporal and spatial association of diverse tones.

Complexes which are essentially homogeneous are usually called *classes*. This is to speak of classes in extension, but it is, as we shall see later,[1] the only legitimate way in which we may talk about classes. In order to have a class there must be some aspect of homogeneity, but in order that the class be a complex there must be some aspect of heterogeneity. Thus a class may be a historical object at a place where the heterogeneity is contributed by the time, or an extended spatial object at a given time where the heterogeneity is contributed by the space, or a limited spatial object at a given time where the heterogeneity is contributed by the qualitative diversity of the associated elements, or any combination of these. If we pass to a sufficiently high level of abstraction every complex may be shown to be homogeneous; hence every complex is a class. The most common case of a class is a temporally homogeneous but spatially heterogeneous dissociative complex whose elements are qualitatively alike in a relatively concrete way. If such a complex is loose-knit, it is usually called a group, pile, aggregate, heap, conglomerate, mixture, or collection; if it is highly integrated, it is usually called an organized or organic whole, a totality, compound, structure, or system.[2]

[1] Chapter VII, p. 155
[2] Cf. Carnap, *Der logische Aufbau der Welt*, p. 48.

The Data of Science

If the task of science is the exploration and symbolic characterization of the realm of occurrents, then the preliminaries to our analysis of the logical structure of science have been properly introduced; we have pointed out the essential kinds of elements and the important types of structure and organization which are to be found in this realm. We need only indicate now how science proceeds with its task.

But is it the task of science to explore and characterize the total realm of the given? Or is science to be limited for its subject-matter to some portion of that realm? For example, it is usually stated rather glibly that science is concerned with the study of " nature." Now if this word is used to designate the totality of occurrents, i.e. the totality of things " given," the statement need not be called into question. But the word is so often used as a cover under which to smuggle in a more limited interpretation. By nature may be meant not the totality of things experienced but the " natural " as opposed to the " supernatural," or that which is " real " in nature as opposed to that which is " unreal," or that which is " outside " the mind as opposed to that which is " inside " the mind, or that which is " concrete " as opposed to that which is " abstract."

I am unable to see the justification for this limitation, at least as a principle to be employed anterior to science for the purpose of determining what the legitimate data of science are. It may be that after science has completed its task it is able to classify its data into these various realms. But prior to science such classification cannot be made. The reason for this is that the actual determination of the realm in which an occurrent lies is precisely science itself. The location of an occurrent in its proper realm is a matter of discovering its qualities and relations; but the discovery of the characters and connections of an occurrent constitutes science. As Ritchie points out,[1] prior to science white blood corpuscles and pink rats are upon precisely the same level. But by subjecting both of them to scientific study we discover that the former exhibit certain properties which the latter seem to lack; we then put the latter upon the scientific

[1] *Scientific Method*, pp. 28–30.

waiting-list pending more information as to their behavior. Clearly, if the characters determining the location of an occurrent are not given with the occurrent they are not given at all, for there is no other place where they may be given. And it seems unreasonable to suppose that science must be postponed until there has been a critical study whose aim is the division of its objects into the proper metaphysical realms.

It follows, therefore, that science must start with pure phenomenalism. If the subject-matter of science is nature, then nature must be defined as synonymous with the *given*—according to the most generous meaning of this word. Then the subject-matter of science is the realm of occurrents—both those which are given directly and those which are given referentially, (i.e., through symbols); both those which are given in the present and those which have been given and will be given; both those which are given here and those which are given there; both those which obey natural laws and those which do not; both those which exhibit the properties of objectivity, publicness, and permanence, and those which do not; both those which are material as to stuff and those which are not; both those which are given concretely and those which are given abstractly; and so on. The subject-matter of science is, in short, the totality of things actual and possible.

But the fact that science begins with phenomenalism does not mean that science remains forever phenomenalistic. It is precisely the task of science to replace the realm of the unclassified and unordered given, by a classified and ordered given, through the application of a principle which will determine the selection of data. This principle is two-fold : (1) With reference to the realm of the given, that which is given clearly and distinctly should always be subjected to examination prior to that which is given obscurely and indistinctly. (2) That which is given obscurely and indistinctly should be defined through its relationship to that which is given clearly and distinctly. It will be the task of the following chapters to elucidate this principle.

For the present we shall be concerned merely with the application of this principle to the determination of realms of the given. As a principle of selection it seems to offer just such a methodological criterion as is demanded. In approaching the study of the given *something* must be neglected or at least set aside; attention is relatively narrow in its range and we cannot gain a distinct awareness of *everything*. What, then, shall we relegate to the waiting-list? With reference to the distinctions listed above, however they may be defined, there seems to be no great difficulty in determining in each case which of the alternatives is *clearly* given and which is *obscurely* given. Certainly those occurrents which are given directly are given more clearly than those which are given referentially, and those which are given as non-symbolic are given more clearly than those which are given as symbolic (i.e. roughly, things are more clearly given than symbols for things). Again, occurrents which are given in the present are given more clearly than those which are given in the past or the future, and those which are given here are given more clearly than those which are given there. Occurrents which obey natural laws are given more clearly than those which do not . Occurrents which exhibit objectivity, publicness, and constancy are given more clearly than those which do not. Occurrents which are " observed " seem to be given more clearly than those which are " introspected "—if by " clearness " is meant not " certainty of existence " but " definiteness of character." Finally, as we have already seen, the " middle-sized " occurrents are more clearly given than either the high concretions and the high abstractions, or the highly extensive and the highly minute.

What is the result? The data of science become divided into two groups, the *clearly given* and the *obscurely given*.[1] (Actually most of this division has been done tentatively by common sense prior to the advent of science as a specialized study, but it is a selective process which is of the same kind as

[1] The recognition of some such dichotomy as this seems to pervade the literature of the philosophy of science. So far as I am able to ascertain, the distinction between the *clear* and the *obscure* expresses in a more empirical way the following equivalent distinctions: H. Poincaré's " crude " fact and " scientific " fact (*Foundations of Science* (New York, 1921), pp. 325-333); E. Meyerson's " law " as empirical rule and " law " as underlying identity (*Identity and Reality* (New York, 1930), chap. I); Jeans' " shadow " world and " mathematical " world (*op. cit.*, chap. V); Eddington's " pointer readings " and " unscrutable counterpart " (*Nature of the Physical World*

science and therefore continuous with science). On the one hand are the clearly given; these are the occurrents which are directly given, non-symbolic, present, here, natural, real, observable, and " middle-sized." All other data are obscurely given with reference to one or more of these criteria—which are not assumed, of course, to be mutually exclusive. Obviously science should begin with those occurrents which show the greatest promise, —viz., the clearly given occurrents. Hence, the first task of science is that of attempting to determine the characters and inter-relations of these clearly given occurrents. During this investigation the latter group of data is set aside, not because it has a position which is metaphysically inferior, but because it is inherently vague; i.e. we cannot understand it in itself and we have not yet acquired anything else in terms of which it can be understood.

But now let us suppose that we have determined the properties and relations of the clearly given realm. We are then ready to turn to the other realm. How is this to be understood? There seems no alternative to supposing that what we do not know must be explained in terms of what we do know.[1] The obscure must be made clear in terms of the only available instrument,— viz., the data which have already been classified and ordered. But all of this presumes that the realm of obscure data has significant relations to the realm of clearly given data. More specifically it presumes that every obscurely given datum is united in a relational-complex with at least one clearly given datum, and also, if knowledge is indefinitely expansible, that every clearly given datum is united in relational-complexes with a large number of obscurely given data. Then the growth of knowledge consists in the progressive revelation of more and more relational-complexes in which any clearly given occurrent participates, and, conversely, in the progressive " reduction "

(New York, 1929), p. 254); M. Planck's " world of our senses " and " real world "—which latter world is known only through the " world of physics " (*Universe in the Light of Modern Physics* (New York, 1931), chap. I); Whitehead's " sense objects " and " scientific objects " (*Concept of Nature*, p. 149); Russell's " hard " data and " soft " data (*Our Knowledge of the External World*, pp. 70–73); Stace's " factual existence " and " existential construction " (*op. cit.*, chap. VII); Carnap's " *erkenntnismässig primär* " and " *erkenntnismässig sekundär* " (*op. cit.*, p. 74).

[1] Cf. Stace, *op. cit.*, p. 380.

of obscurely given occurrents to the clearly given occurrents with which they are united in relational-complexes. For example, if we suppose a small lump of gold to be a relatively clearly given occurrent, then knowledge advances when we learn more about the gold, e.g., its history, its properties, and its constituents. But its history can be determined only by its present condition; its abstract properties have meaning only in relation to the concrete properties given in the gold, and its microscopic aspects must be given content in terms of the macroscopic features.

It follows that the realm of obscurely given occurrents becomes differentiable into sub-realms upon the basis of the general kind of relation connecting the obscure occurrent with the clear one. Those occurrents which are related temporally are placed in the realms of the past and the future; those which are related through spatial distance and direction are located within the realm of the remote spatial. Those which are related through spatial and temporal inclusion are located within the realms of the highly extensive and the highly minute. Those which are related through association are located within the realms of the highly abstract and the highly concrete. Those which do not have significant spatial or temporal relations yet are intimately related through an " observer " become located in the various realms of symbols, hallucinatory and imaginative objects, fictions, etc.

The use of the word " realms " should not confuse the issue. It should not be taken to mean that these realms have location at a time or at a place, though, since they all have more or less precise spatial and temporal relations with the realm of the clearly given, there is reason to believe that they could all be located in that realm. But the task seems unimportant in many cases. For example, probably all imaginative entities have residence in or near the brains of those who are thinking them; but the important thing about an imaginative entity is not where it is but the kind of situation, consisting of a clearly given datum and an observer, in which the imaginative entity arises and the way in which its content is determined accordingly.

The Task of Science

The task of science is now determined. It is the ascertainment of the relations of the various obscurely given occurrents to the realm of the clearly given.[1] This is a problem of great complexity, for there are many types of relation uniting obscurely given occurrents to the clearly given. Hence the problem reduces to many more specific problems, each more or less independent of the others. Each of these is a problem of the determination of the content of an obscurely given occurrent by virtue of its relation to a clearly given occurrent. Hence, there arise the several problems of the determination of the *meaning* of the past and the future, the *meaning* of the remote spatial, the *meaning* of the highly abstract and highly concrete, the *meaning* of the highly extensive and highly minute, the *meaning* of symbols, the *meaning* of the supernatural, the unreal, the hallucinatory, the introspective, etc.—all upon the basis of the examination of the realm of the clearly given. Apart from some connection with the realm of the clearly given, these realms are meaningless. The only alternative to the "reduction" to the more obvious occurrents is to continue to think about the obscure occurrents in a vague and indefinite way, in the hope that clarification will ultimately come. But such clarification, when it comes, must consist precisely in this revelation of the relation of the obscure occurrent to some clearly given occurrent. Hence it seems more profitable to undertake immediately to find the significant relation.

But a most important difficulty in the problem of the ascertainment of the relations of obscure occurrents to clearly given occurrents must now be pointed out. It is the fact that the relations of the obscure occurrents to the clearly given occurrents are themselves usually given only obscurely. Thus it becomes an extremely important problem to determine what these relations are. If these relations were given clearly, the task of giving meaning to the *relata* of the relations would be simple. We should be obliged merely to start from the clearly given occurrent, note its relations, and follow out these relations until we reached the terms thus connected to the datum; these

[1] Cf. Carnap, *op. cit.*, pp. 252–5.

terms could then be given meaning directly in terms of their relational connections with the realm of the clearly given. We should have the tentacles from the real which would enable us to locate the obscure and thus bring it within the compass of the clearly given. But such is not the case. It is not definitely given that every occurrent has a past, or a future, or a symbol, or abstract properties, or larger spatial and temporal wholes within which it is included, or smaller spatial and temporal elements which it includes, or hallucinatory and illusory forms, etc. The only thing to guide us here is the general uniformity of nature; most occurrents have all of these relational connections; hence any given occurrent may be presumed to possess them.

However, the situation is not actually so bad as this. Although obscure occurrents do not always reveal their *relations* to the clearly given, they do usually reveal the *operations* through which they were obtained and through which their meaning is to be ascertained. Since the question of operations is to be postponed to a later chapter,[1] our discussion must here be brief. The point is that although we cannot often see how an obscure occurrent is related to the given, we frequently know how we got it.[2] We recall that when we performed a certain operation upon a given occurrent we suddenly found ourselves aware of the obscure occurrent. We can therefore define the obscure occurrent at least in the sense that we can show where and how we got it. Thus I can say that a white blood corpuscle is what I see when I put blood under a microscope, that the events of yesterday are what I become aware of when I start with the events of today and begin to remember, that the circularity of a penny is what I experience when I look at a penny and perform an abstraction on it, that society is what I get when I think of individuals and then think of them as being united by social ties, etc., etc.

But now a strange thing happens. Since the operation suffices both for the purpose of locating the obscure occurrent and for the purpose of giving it precision, the question arises

[1] Chapter VI.
[2] Cf. C. I. Lewis, *Survey of Symbolic Logic* (Berkeley, 1918), " an operation is something *done, performed,*" p. 342.

as to whether the relation at the foundation of the operation is not superfluous. If the operation will do all that was demanded of the relation, why search further for the relation? In fact, why suppose a relation at all? And if there is no relation, why suppose that the result of the operation has any metaphysical status other than its constructibility through an operation? All obscure occurrents then become constructs, devices, inventions—derivable from the clearly given but not a part of it. This has seemed to many to result in a wonderful simplification of reality,[1] for at one stroke realms and realms of entities have been eliminated from the world. The past and future, remote space, abstractions, highly extended and very minute occurrents, symbols, hallucinations, imaginative entities—have all been removed from the realm of existence.

The full criticism of this position would take us too far afield. But its essential difficulty seems to lie precisely in the fact that when the operation becomes public and repeatable and always results in the same construct, there is no longer any criterion for distinguishing constructs from clearly given occurrents. When an operation is performed repeatedly it becomes unconscious, and we tend to believe that the resultant entity is a discovery, not an invention. Thus most of the clearly given entities can be shown to have been at one time obscure entities obtainable only through operations. Abstraction, analysis, synthesis, classification, inference, and negation are some of the most common operations which soon take on an unconscious status, and we thus are led to *find* abstractions, elements, wholes, classes, implicates, negatives, within the realm of the clearly given. Consequently, there is no way of telling precisely what a construct is, as differentiated from a clearly given occurrent, and we are not entitled to place all operational entities in a realm of the unreal.

Furthermore, the positivist of this school makes an important distinction between such constructs as atoms, molecules, and white blood corpuscles, on the one hand, and fairies, centaurs and mermaids, on the other. The latter group have even less

[1] E. Mach, *Popular Scientific Lectures* (Chicago, 1910), lect. IX; K. Pearson, *Grammar of Science*, (3rd ed., London, 1911), chap. II; Russell, *Our Knowledge of the External World*, lect. IV.

justification for status within the real. But the reason for this seems to lie in the fact that the operations which are employed in the derivation of these latter entities are not essentially repeatable. And they are not repeatable simply because the distinctive features of the observer are an important part of the operational set-up. All operations may be expressed with the operator as a part of the picture. Thus one may say that the past is what someone remembers, and the future is what someone expects; so also abstractions are what someone gets by abstracting, etc. But one feels that it is important to mention the observer only when different observers would bring about different results. If the operation is essentially repeatable with essentially the same result we may neglect the observer. Now we find that we may define the past as what has been, regardless of who remembers, since in general we agree reasonably well as to what has happened. So also the future is what will be, no matter who expects it; the remote is where anyone may go; the abstract is what is abstractly present in any situation; a symbol is what one normally thinks of when he becomes aware of a thing. But hallucinatory objects, imaginative entities, objects of introspection, etc., seem to vary markedly from individual to individual, hence we tend to feel that the observer is an important part of the total situation and that the operations are consequently not repeatable and public. We are therefore not tempted to put the derived entities into the realm of the clearly given. We place them aside while we await a more detailed analysis of the operation. But this means that all operationally derived entities are of essentially the same kind since the fact of the repeatability and the publicness of an operation does not give it a privileged metaphysical position. It is not too much to expect that just as atoms, molecules, and white blood corpuscles pass from the realm of constructs into the realm of the clearly given through a generalization of operational techniques, so also fairies, mermaids, and centaurs will be subjected to the same change of location when the physiological and psychological operations involved in their creation are better understood.

Let us return to the matter of operations. We are required

to define the obscurely given occurrent through its relation to the realm of the clearly given, but we find that this relation is itself given only obscurely. Hence we replace the relation by the operation which is presumably its equivalent. We now have available all of the material in terms of which the obscure occurrent is to be defined. But since we wish to talk about this occurrent in advance of any clear awareness of it, we must replace the occurrent by a symbol for it; when an entity is not given clearly, we replace it by something which acts as its substitute and yet may be given clearly. Hence our problem is now to define the symbol for the obscure occurrent through the proper operational route. Immediately the suggestion arises as to whether we cannot define the symbol by that ultimately satisfactory method of definition, viz., the denotative method, or the method of pointing. But the denotative method necessarily fails in the case of obscure occurrents; they are obscure precisely because we cannot unambiguously point to them. Yet the ultimate meaning of our symbol must be found in some sort of denotative reference. If we cannot point to the occurrent itself, we can perhaps point to another occurrent which is clearly given and with which the obscure occurent has known relations. This is the only possible method for locating it in the realm of the given; it becomes " reduced " to the given by virtue of its intimate union with the given. Hence, the symbol for the obscure occurrent becomes definable through this route of location and through the character of that with which it is located. Since the actual relation uniting the obscure occurrent with the clear occurrent is not unmistakably given, it is replaced by its operational formulation. We then define the symbol for the obscure occurrent in terms of the operational route which must be followed out by an observer in passing from the clearly given occurrent to the obscurely given occurrent. This gives the symbol for the obscure occurrent content in the only way in which it can be given such content; it employs the denotative method by directing the observer to a situation to which something must be done before he can experience the occurrent in question. Various directional techniques define various routes of location, and the routes of location constitute, in advance of any clear awareness, the only method for giving content to the obscure occurrents.

There is a slight additional complication in the designation of this operational route. Since we cannot always "get" even the clear occurrents conveniently, we replace them by symbols for them. The denotative method is sometimes, at the moment, inapplicable; hence we must rely upon the connotative method. Occurrents given directly are always preferable to occurrents given through symbol, and the ultimate reference of the symbol must be to occurrents; but in many situations we find it convenient to use the symbol as a substitute for the occurrent. Hence, the complete operational reduction of obscure occurrents can be expressed by asserting that obscure occurrents are replaced by symbols for them, and these symbols are defined by their symbolic relations (defined in terms of operations) to other symbols which refer directly to clear occurrents. Thus there is the symbol for the obscure occurrent, the symbol for the relation, and the symbol for the clear occurrent. Then the symbol for the obscure occurrent is given content in the only manner possible,—viz., through the symbol for the relation and the symbol for the clear occurrent. This constitutes the operational definition of the symbol for the obscure occurrent.

This means that symbols in this function lie midway between the most clearly given and the most obscurely given. When we have both *clearly given occurrents* and *symbols for such occurrents*, we always forsake the symbols in favor of the occurrents. But when we have both *symbols for obscurely given occurrents* and *obscurely given occurrents*, we always forsake the latter in favor of the symbols, for the symbols are more precise as to content than are the occurrents. Thus if we know more about occurrents than we do about symbols, we explain the symbols by referring to the occurrents; but if we know more about the symbols than we do about the occurrents (as is often the case), we explain the occurrents by referring to the symbols. Symbols are inadequate substitutes for occurrents when the occurrents are clearly given, but they are adequate substitutes when the occurrents are obscurely given. Hence the secondary task of science is that of giving obscure occurrents precise content by " reducing" them to clearly given occurrents, the "reduction" taking place through the intermediation of symbols. The obscure occurrents are then simply what we become aware of when we perform certain operations. That they do not remain thus narrowly defined is due to the fact that as we perform these operations repeatedly the obscure occurrents become less obscure; by habituation we discern novel features, so that the obscure passes imperceptibly over into the clearly given. And we gradually become capable of grasping the occurrent in awareness without the intermediary operation. This must suffice for the preliminary statement of a topic to be developed in greater detail in the pages which follow.

CHAPTER V
AWARENESS

We pass now from metaphysics to logic. The foregoing excursion into problems of nature was found to be necessary because of the character of the scientific enterprise. If science is the exploration of nature and the formulation of her content and structure in symbols, some examination of the realm of natural entities is a prerequisite to undertaking a logical analysis of scientific method and doctrine. This examination has been completed, and we are ready for the more detailed examination of awareness.

Let us recall those features of awareness which we discerned in our discussion of the general character of scientific cognition. The most important conclusion to which we came was that awareness is not knowledge unless it is the awareness of a complex content consisting of an occurrent, a symbol for the occurrent, a symbol for the applicability of this symbol to the occurrent, and a symbol for the justification of this applicability. This, at least, is *adequate* knowledge; and the absence of certain portions of this complex content results in different kinds and degrees of *inadequate* knowledge. It follows that mere awareness is not in general to be identified with knowledge though it does constitute the essential feature of knowledge. We shall do well, therefore, to begin with an examination of awareness.

Relation of Awareness to its Content

Awareness has been shown to possess a peculiar relation to its content which permits us to say that it is *of* the content. There is probably very little that one can say which will elucidate further the character of this relation. It is perhaps one of the indefinables of knowledge. But one of its features has already been suggested, and we turn to a more detailed examination of this property.

However intimately the awareness may be related to its content, the latter must be in some sense independent of the former. This feature of the content was called in our earlier discussion its " objectivity," and it was judged to be a necessary attribute of the content if the latter is to be rational or knowable. This means, crudely stated, that awareness does not transform its content. The only demonstration of this is through the *reductio ad absurdum*. If we assume that awareness does transform its content, there are two consequences. We can never have genuine knowledge of reality, and we can never *know* that awareness does transform. The former follows from the fact that our knowledge must always be of a *transformed* reality; and this, so far as our knowledge is concerned, must be the *only* reality. If we assume an " external " reality which is transformed by the act of awareness, then we are entitled to ask what the character of this reality is. But the question is clearly incapable of being answered, and an assumption which makes the question unnecessary seems preferable. The latter consequence,—viz., that we can never know that awareness does transform—follows because to demonstrate the transforming power of awareness we should be obliged to compare the content as transformed with the content as untransformed. But this would presuppose another awareness which did not transform; hence we should be obliged to admit at least one awareness which is nontransforming. Then we should have the difficult problem of determining which awareness transforms and which does not. It seems simpler to assume that no awareness transforms.

But is not this contradicted by the facts? Observation, for example, quite obviously transforms. One sees a straight stick immersed in a glass of water as bent. Here awareness transforms, for what is observed is no what exists. Again, thought transforms, because it is in many cases erroneous. One asserts a proposition which proves to be false and is thus a transformation of its object. Again, fictional operations resulting in the construction of perfect levers, ideal geometric figures, frictionless motion, etc., produce transformed objects so that what enters into awareness is not what exists.

The error here involved is in the choice of examples. When one thinks of awareness as transforming, he usually calls to mind such activities as observation, thought, and imagination.

But all of these are relatively complex forms of awareness, and each of them might conceivably involve two or more simpler awarenesses of distinct contents. This seems to be the explanation of the difficulty. It is possible in complex forms of knowing to be aware of a given content and to be at the same time aware of another content having a certain relation to it. Thus one can be directly aware of two contents, differing from one another but of such character that one has been derived from the other by an operation. It appears, therefore, that awareness has transformed. But there seems no reason to suppose that *awareness* has transformed; one need only suppose that there has been a transforming act which has created a new content for awareness. What one is aware of in any case is simply what is given for awareness; the question as to how this entity given for awareness is produced is independent and must be answered independently. It is clear that in all of the cases here considered there is something given directly for awareness, and that awareness does not transform this entity. The stick is given as bent, the proposition as possessing a certain content, the lever as perfect, etc. In case of apparent transfomation, then, it is necessary only to show how the transformation has been accomplished and how a new entity has been created for awareness. This means that in most ordinary cases of awareness there is a complex situation which involves the related awarenesses of related contents; this, as we shall see, is not simple awareness but knowledge.

The fact that awareness is not transforming may be expressed in the equivalent proposition that the addition of the awareness to the content does not necessitate any change in that content. The relation of a content to awareness is an external relation. It may be convenient, however, to designate the fact that a content is entering into awareness. We may therefore introduce the notion of *idea* to designate a content of which someone is aware. The content may be called an idea only when and so long as it is within the awareness of an individual. When a content becomes an idea, it does not take on any different metaphysical status; it does not become mental, or in consciousness, or in the brain, or in private space and time. All of these conceptions are the result of the erroneous theory of knowledge associated with English empiricism. It is false to say that we are aware only of ideas ; we are aware rather of contents, which by virtue of their being objects of awareness are also characterizable as ideas. To say that one becomes aware of an idea is like saying that a man becomes married to a wife. One should say rather that a man marries a woman who, by virtue of the marriage, becomes a wife. One could become aware of an idea if one meant that he became aware of a content of which someone else was at the same time aware. Thus one could also say that one married a wife (of another man). But this is not the usual sense in which one speaks of becoming aware only of ideas. Furthermore, error must not be looked upon as a discrepancy between idea and occurrent. There can never be any such discrepancy, so far as the contents of the two are concerned. Though they are distinguishable, they are distinguishable only by the fact that the latter is increased by something in order to produce the former. To suppose that unreal objects exist only as ideas is to introduce confusion. There is no reason to suppose that a centaur of which I am aware has the same kind of existence as have ideas, unless ideas are put into a realm of unreality. If ideas are occurrents which are entering into awareness, then a centaur is an idea when I am aware of it and it is not an idea when no one is aware of it. An idea could never occur in the absence of its occurrent, for it *is* this occurrent in a certain relationship. Finally, an idea must not be identified with a symbol ; a symbol will be an idea if anyone happens to be aware of it, otherwise it will not be. Thus the distinction between a symbol and its referent is not to be identified with the distinction between an idea and the occurrent of which it is the idea ; an idea could never be found apart from its occurrent, but a symbol can be and often is found apart from its referent. The fact that an occurrent may exist apart from its idea and a referent may exist apart from its symbol should not lead to a belief that the two situations are completely analogous.

The word " idea " as thus defined is one of those peculiar philosophical words which, once defined and introduced into a discussion, may be dismissed and not referred to again. The only occasion for the employment of the word would be a situation in which an individual is aware of a content,—e.g., an occurrent. We should then use the word to designate an occurrent which we *have* as opposed to an occurrent which we *do not have*,— i.e., " have " in the only sense possible—through awareness. But it we *have* the occurrent it is unnecessary to speak of it as an idea since the fact of its being an idea makes no difference to it. And if we *do not have* the occurrent at all through awareness, then we must have it in some indirect sense. But this indirect sense is very important and is specifically defined; viz., it is *having through a meaningful symbol*. Thus when the occurrent is given for awareness, we *need not* speak of it as an idea; and when it is not given for awareness, we *cannot* speak of it as an idea but must speak of it as the referent of a symbol. Hence the word " idea " becomes unnecessary.

Intensity of Awareness

We shall now point out the three important ways in which awareness seems to be under control. The first is *intensity*. We are clearly aware or only vaguely aware. Vague awareness passes imperceptibly into non-awareness; often we cannot be sure whether we are aware or not. It may be that differences in intensity of awareness are simply differences in distinctness of occurrents; i.e. it may be that when we say that we are vaguely aware, we are aware of a vague occurrent; and that when we are distinctly aware, we are aware of a distinct occurrent. This is a plausible hypothesis. It seems better, however, to suppose that all occurrents are distinct, and that vagueness is a weakness of human knowledge. Attention is not therefore an operation by which we pass from the awareness of a confused occurrent to the awareness of a distinct occurrent. Rather one should say that attention is the activity by which a confused awareness of an occurrent is replaced by a distinct awareness of the same occurrent, though we can never know that it is the same occurrent. Through control we can replace a vague awareness by a distinct awareness; at least we can produce such distinctness as the physical situation permits. In general, a distinct awareness is to be preferred to a vague awareness, for it is only through a distinct awareness that we can be relatively certain of the applicability of a symbol to an occurrent.

Kinds of Awareness

Secondly, awareness is under control, at least partly, as to *kinds*. That which determines the kind of awareness is the kind of entity which enters into the awareness. This has a certain consonance with common sense; memory, anticipation, observation, imagination, introspection, thought, are usually considered to be different kinds of ways by which objects may be made known. I shall attempt to make precise this notion of kinds of awareness. Since everything which we know must be an occurrent, we may say that all awareness is of occurrents. But occurrents, as we have seen, are of many kinds: symbolic and non-symbolic, present, past and future, natural and supernatural, real and unreal, material and psychical, concrete and abstract, etc. If the character of the awareness is determined by the character of the entity which enters into awareness, we shall be obliged to assume as many kinds of awareness as there are kinds of content. I can see no significant objection to this, except that in many cases, since the distinctions between the types of content are not sharply drawn, there would result a corresponding obscurity in the distinctions between kinds of awareness. Probably if one searched through philosophical literature, he could find a term to designate the kind of awareness associated with each of these types of element. For our purposes only those distinctions which are clear-cut need be emphasized.

One of the most obvious of these distinctions, and that about which we are least often mistaken, is that between symbolic and non-symbolic occurrents. A symbol possesses a meaning property by which it becomes a new type of entity having a different form and a different content. Thus one may make a general distinction between awareness of symbolic occurrents and the awareness of non-symbolic (or ordinary) occurrents. In the former case the awareness is essentially of the occurrent itself, whereas in the latter case it is primarily of the meaning of the occurrent and only secondarily of the occurrent itself. Let us employ the word *thought* to designate the awareness of a symbolic occurrent. Then thought is the awareness of words, pictures, images, models, diagrams, indices, gestures,—of any occurrent, in fact, which has the capacity to mean or point in a cognitive sense to something outside of itself. Thought may or may not be associated with an awareness of that which is referred to by the symbol; if there is this additional awareness, then the situation should no longer be described simply as thought. But the fact of this additional awareness introduces confusion into the use of the term. For when one says that he is thinking of an occurrent, he must mean that he is indirectly aware through an occurrent which is a symbol, of an occurrent which is not a symbol. He is thus apparently aware of two occurrents—a non-symbolic occurrent indirectly, and a symbolic occurrent directly. Actually, it is only the symbolic occurrent of which he is aware; the non-symbolic occurrent is given only through the meaning of the symbol. Hence it is more precise to say that one *thinks* symbolic occurrents and thus *means* non-symbolic occurrents.

A particularly interesting kind of thought is the awareness of symbols which have the capacity to refer to the past and to the future. Our knowledge of the past and future must always be indirect; we cannot be aware of the past or of the future, but we can be aware of occurrents which have the capacity to represent them. This symbolism may be either of the pictorial character, in which case we form images of the past and future, or of the linguisitc character, in which case we use words to describe them. Thus *memory* may be defined as awareness of symbols for the past, and *expectation* as awareness of symbols for the future.

Contrasted with thought, there is the awareness of an occurrent as non-symbolic. This form of awareness has no specific name, since it includes many more specific kinds of awareness,—e.g., observation, introspecton, fancy. It is any awareness in which the content of the awareness is something which may be referred to rather than something which is referring. To be sure, a given content may be both that which is referred to and that which refers; e.g., the word " cat " may be both that which is symbolized by the French " chat " and that which refers to an actual cat, but it cannot be both in the same sense. More precisely, if we know it as the referent of " chat " we are aware of it as non-symbolic; hence we do not think it. If we know it as that which refers to an actual cat, we are aware of it as symbolic ; hence we do think it.

The use of the more specific terms to designate the various kinds of awareness of non-symbolic contents will reflect the difficulty which we experience in determining the metaphysical status of an occurrent. The distinctions between the natural and the supernatural, the real and the unreal, the material and the psychical, the abstract and the concrete, the extended and the minute, etc., are not capable of being so sharply drawn as is the distinction between the symbolic and the non-symbolic. For example, the distinction between the abstract and the concrete and the distinction between the extended and the minute, are both purely relative in character; hence we cannot specify kinds of awareness associated with them. Attempts have been made along the lines of rationalism to designate the awareness of abstractions as " reason " and the awareness of concretions as " sense-perception." This only confuses the issue, for it suggests that the distinction between abstract and concrete is identical with the distinctions between universal and particular, and between symbol and referent—and neither of these views seems to me to be correct. Many of the above distinctions (e.g. that between the natural and the supernatural, and that between the material and the psychical) seem to be disappearing from philosophic thought. The distinction between the real and the unreal seems to forbid accurate formulation.[1] While we are dreaming we cannot know that we are dreaming, and while we are observing we cannot be certain that we are observing. For practical purposes, as we saw in the preceding chapter, we may often distinguish unreal from real occurrents by an examination of the operational techniques employed in becoming aware of them ; as these operational routes become public and repeatable, we tend to suppose that the entities derived through them are real rather than unreal. Since we shall have occasion to refer to this topic later, it will be convenient henceforth to use the terms *observation* and *perception* to designate the process of becoming aware of a real occurrent, and *fancy* and *imagination* to designate

[1] Peirce, *Collected Papers* (Harvard University, Vol. II, 1932), prop. 2.337 ; also Whitehead, *Adventures of Ideas* (New York, 1933), p. 270.

the process of becoming aware of an unreal occurrent. (This use of the term "imagination" must be distinguished from its employment in the sense of "having images" when the images purport to have a descriptive reference; imagination in this latter sense is a kind of thought, since it is an attempt to symbolize occurrents. The imagination which is synonymous with fancy is that which is the uncontrolled succession of contents illustrated by day-dreaming where the contents are taken at their face value).

We may summarize the discussion of kinds of awareness as follows : Awareness is always of occurrents. If the occurrents are symbolic, the awareness is called *thought*, of which *memory* and *expectation* are interesting species. If the occurrents are non-symbolic, the awareness has no specific name, though it is differentiable into many kinds, corresponding to the various types of occurrent which may enter into awareness. One of the most useful of these distinctions is that between *observation*, which is the awareness of real occurrents, and *fancy*, which is the awareness of unreal occurrents. It is impossible to ascertain the kind of awareness in which we are engaged until we know the kind of occurrent of which we are aware. This is an expression of the fundamental dependence of awareness upon its content.

Locus of Awareness

Thirdly, awareness is under control as to its *locus*. This fact is usually expressed by saying that awareness is selective toward the realm of occurrents. The sense in which this is true, and the degree to which it is true even in this sense, are both hard to explain. Questions must be raised here which can be much better answered in the next chapter after we have completed the discussion of operations. But it can be indicated, I think, that if selection is possible, it must be because the realm from which the selecton is made has a certain definiteness and independence. Conversely, if the realm from which selection is made is indefinite or obscure and essentially dependent or referential in character, selection is impossible simply because in such a case it becomes indistinguishable from invention. The criteria of inventions as over against discoveries, as we shall see in the next chapter, are the relative obscurity of invented notions as compared with the entities from which they are derived, and the fact of the inherent reference of the invented entities back to the situations from which they were obtained.

It follows that the selectivity of awareness must be determined not for occurrents in general but for certain realms of occurrents. Only the realm of the clearly given occurrents permits of strict selection. Here the determinateness and independence of the realm enable us to explore its content. We *discover* what is there, since the occurrents are given clearly and awareness is passive toward its content. Given such a situation, we may change the locus of awareness in accordance with changing interests and in accordance with the structural features of the situation itself. We may attend to the situation as a whole (provided it is not too extensive), or to any spatial part, or to other situations connected in spatial ways, or to co-associates of the situation, or to slightly more concrete or slightly more abstract associates. Here the operations concerned with the change of locus are purely exploratory, and awareness may therefore be said to be selective.

It is clear that in such cases awareness is subject to limitations ; it must remain within the realm of the clearly given, and can therefore direct itself only upon that which is " there " for awareness. We cannot, for example, select any occurrents from the past, and we cannot select any occurrents from the future unless we are willing to wait until they occur in the present. We cannot become aware of occurrents remote in space unless we are willing to go to the proper location. We cannot become aware of high concretions or high abstractions, or of the very large or the very small, or of that which is so situated physically as to prevent awareness,—e.g., the other side of the moon or the inside of an opaque body. Thus the freedom of selection is a freedom within a designated realm only.

As we pass to the realm of obscurely given occurrents, selection becomes impossible. If occurrents are not such as to enter clearly into any awareness, it is absurd to talk of making a selection from among them. The situation is very much complicated by the fact that, although we cannot readily become aware of the obscure occurrents, we can and do very readily become aware of symbols (usually images) for them. Thus apparently we can call them to mind at will. But three points should be mentioned here. In the first place, it is the symbols and not the occurrents which are called into awareness. In the second place, the symbols are not selected (discovered) but invented; this is true especially of those symbols which are most easily conjured up in the mind,—viz., images. In the third place, if the symbols appear to have a certain definiteness and independence of content, as is the case with many word symbols, then by this very token they ought not to be located in the realm of the obscurely given. Word-meanings are not purely arbitrary but are socially determined. To the extent that they are socially determined they are socially fixed and therefore socially discoverable. Hence they belong in the realm of clearly given entities, and can be explored exactly as can the rest of that realm.

A summary of these results might be possible in somewhat the following terms : In the broadest sense awareness is selective toward its content; we usually express this by saying that we can " think " of anything we please, and we can change the content of our "thoughts" at will. But this does not mean that we can be aware of any occurrent we choose to call to mind,—though it does mean that if an occurrent cannot be produced in awareness, we may easily substitute for it a symbol which will act in its stead. The only occurrents which may be produced in awareness are the occurrents which are there to be discovered. But if for any one of a number of reasons we cannot become aware of an occurrent, we may invent a symbol for it. This serves practically as a very valuable substitute. But it is dangerous, because the symbol is an invention. This means that it is indeterminate and referential as to content. The indeterminateness is generally recognized, but the referential character is usually forgotten. Hence, the symbol is made determinate not by reference to the realm of occurrents, as the operational theory demands, but by the free activity of " imagination." This contributes both to vagueness in thinking and to the supposition of all sorts of entities not demanded by the empirical situation. The only way to avoid this danger is to recognize frankly that symbols ought not to be used as substitutes for occurrents unless we recognize clearly the regulative principles which must be employed in determining their meanings. These principles are founded upon operational acts, and it is to this phase of the subject that we shall turn in the next chapter.

Observation

Accepting this examination as a preliminary, let us now attempt to make a more detailed analysis of a specific kind of awareness. Any one of a number of forms might be chosen, but I have selected *observation*, since this is the essential form for science. Our knowledge of the general features of awareness thus far discussed will help us in making this analysis. Most analyses of cognition make a mistake in their approach; either they plunge directly into the study of a given situation without any preliminary discussion of the kind of elements to be discovered therein (i.e. they attempt to employ the analytic method without a clear idea of the character of the whole which is to be analyzed[1]), or else they begin with that extremely rare type of situation in which there is an undifferentiated awareness of something—an awareness not of a somewhat, or of this as opposed to that, but a vague consciousness that something is given.[2] Both approaches are pedagogically bad,—the former because of its abruptness and the latter because of its artificiality. I shall attempt, rather, to examine a concrete awareness-situation upon the basis of a recognition of the presence in such a situation of certain elements already distinguished,—viz., the awareness itself; its intensive variability; its selectivity; its dependence upon its content; its non-transforming character; the objectivity publicness, and constancy of the content; etc.

Let us take a concrete situation, such as the observation of a table. We know in advance that the awareness will have that peculiar relation to the table which we describe by saying that it is *of* the table; we know also that the table is presumed to have a certain objectivity, publicness, and constancy, and that

---

[1] G. E. Moore tends to make this mistake. Cf. his *Philosophical Studies*.
[2] F. H. Bradley falls into this error. Cf. *Appearance and Reality* (London, 1893), chap. XIX.

the act of becoming aware of it does not transform it ; we know also that the awareness is under our control and can be stopped, clarified, and changed almost at will. These may all be called the formal aspects of the situation, because they are not concerned with my seeing the table as a table rather than as a chair or a tree.

More specifically, the situation seems to reveal two general aspects, each of which may be analyzed into sub-aspects. The two general aspects are (*a*) the awareness of an occurrent as a table and (*b*) the awareness that the table is real. The former may be generalized into the awareness of an occurrent as of a certain specific kind; the latter may be generalized into the awareness of an occurrent as residing in a certain metaphysical " realm."[1] The former enables us to determine that our occurent is *this* rather than *that*, and *to say so* ; the latter enables us to determine that our occurrent is real rather than unreal, symbolic rather than non-symbolic, material rather than psychical, etc. Specific instances of the former do not determine kinds of awareness, but specific instances of the latter do,—e.g., observation, fancy, thought.

We may dispose of the second of these aspects immediately, since it reduces to the first aspect. The determination of the " realm " in which an occurrent is located is continuous with and exactly the same kind as the determination of its character. To say that an occurrent is real, or symbolic, or abstract, or psychical, or present, is essentially the same kind of thing as to say that it is round or red. In other words the recognition of the status of an occurrent involves the discovery of some feature of its content. Real occurrents are derivable through certain operations which are public and repeatable, and this is a fact about such occurrents. Symbolic occurrents exhibit meaning properties, and this characterizes symbolic occurrents. Abstract occurrents are derivable from concretions by an abstractive operation, and this is a fact about these occurrents. This is all in line with our assertion in chapter IV that it is impossible, prior to science itself, to determine what the subject-matter of science is to be. To speak of these realms as

[1] As this notion was employed in the preceding chapter.

metaphysical should not imply the abandonment of empirical criteria. They are metaphysical only in the sense that the designation of realms involves reference to more general features of occurrents than their shapes or colors. Hence it is often more important to determine the realm in which an occurrent lies than to determine its shape or color. But the discovery of the realm is a scientific enterprise and involves a mere exploration of the content if the occurrent is clearly given, and a specification of the reference of the occurrent to one which is clearly given in case the occurrent is obscurely given. In the latter case the nature of the referential relation determines the realm in which the occurrent is to lie.

Let us now turn to the former aspect of the observational situation, viz. that aspect which it possesses in common with fancy and thought. In this aspect it is simply an instance of the most general type of awareness—the awareness of an occurrent as having a character without any associated awareness of the metaphysical status of the entity entering into awareness. This aspect of awareness itself involves three sub-aspects: (1) the awareness of the occurrent as of a definite content, as *this* rather than *that*; (2) the awareness of the symbol which means an occurrent of the kind given in awareness;[1] and (3) the awareness of another symbol which means that the symbol given is applicable to the occurrent given. This would ordinarily be expressed by saying that I am aware of a particular occurrent, that I know what a table is, and that I am aware that the occurrent is a table. For the purpose of talking about these sub-aspects of the cognitive situation we may designate each by a specific term. The bare awareness of the occurrent may be called *intuition*, the awareness of the symbol may be called *thought* (in accordance with the earlier discussion in this chapter), and the awareness of the participation of the symbol in the occurrent may be called simply the *awareness of applicability*. Let us turn to a more detailed consideration of each of these elements of knowledge.

---

[1] These coincide with Russell's "knowledge by acquaintance" and "knowledge by description," *Mysticism and Logic*, chap X.

Intuition

Intuition is characterizable only with great difficulty, for it seems to have no distinguishable features. It is an awareness, without being an awareness of anything that can be specified; we are aware of an occurrent as having a content, but we are unable to say just what content it has. This difficulty is not due to the fluidity or vagueness of the content; the content is fixed and precise, and could not be other than it is without changing the nature of the occurrent itself. Yet we cannot, within this awareness, say what the content is; for to say what the content is, is to become aware of a symbol which means the occurrent. Philosophers are divided as to whether intuition ever occurs apart from thought; the question is formulated in terms of the possibility of sensation without perception, acquaintance without description, givenness without characterizable content. Unfortunately, those who claim to have such a direct and isolated awareness of an occurrent in its particularity are usually of a highly emotional nature and turn their awareness upon such complex occurrents as God or the universe. It seems reasonable to suppose that if there is an isolable intuition, it is not confined to a specific kind of subject-matter.

But regardless of whether intuition ever occurs apart from thought, it clearly occurs as a distinguishable element in an awareness-situation along with thought. This seems to be all that need be assumed. Clearly, awareness of such a kind is demanded by knowledge. There must be something which we know, and there must be some means of getting into contact with that thing. Now we have one very important instrument through which this may be done. This is the symbol. But the essential feature of the symbol is that it gives us its referent not directly but indirectly. It claims to refer to something; yet it does not contain this thing as part of itself. This demand of the symbol is a double one: it is a claim of empirical derivation and a claim of empirical application. Neither of these can be met unless there is an entity from which the derivation may be made and to which the symbol may be applied. Furthermore, it is not necessary merely that there should *be* such an entity; the entity must also *be given*. It must be possible, apart from the awareness of symbols, to become aware of the referents of symbols. This means simply that the awareness of symbols cannot exhaust the character of knowledge, and that consequently there must be a *distinguishable* aspect of knowledge as a whole which may be called the awareness of the bare occurrent. This aspect may or may not occur in actual *isolation* from other types of awareness, but it must occur in *distinction* from other types. Specifically, this direct awareness is demanded by the origin of knowledge. Knowledge arises out of problems, and we must have some assurance that our difficulty is a genuine one—not merely one of symbols. Hence, that which gives rise to knowledge must be something outside of knowledge— outside in the sense of being independent of it but not out of relation to it. But intuition is demanded also in the termination of knowledge, i.e. in verification. Knowledge is an activity of establishing conjectures. But conjectures can be established only by reference to something which is outside of them; hence there must be an awareness which puts us into contact with the data in terms of which corroboration is accomplished.

Another very important reason for supposing that there is such an awareness as intuition is the demand in any symbolic system for proper names or pointing symbols. There must be a kind of symbol which determines the particular locus of a symbolic system, a symbol which portrays the fact that in considering a symbolic system, say a map, there is an activity expressed by saying "here we are," accompanied by a gesture indicating a point on the map.[1] The fact that a symbolic system cannot be adequately constructed without such symbols seems to argue for an activity of application as over and above an awareness of applicability. There is clearly a difference between the awareness that a symbol means a portion of the realm of occurrents, and the awareness that *this* or *that* is the occurrent which is meant. The simplest explanation of that difference seems to lie in the supposition that there is an awareness which is purely of the occurrent in its particularity, i.e., as it is or is not an exemplification of a given symbol.

It may be that this direct awareness of occurrents is to be

[1] I am indebted to Wittgenstein for this penetrating truth.

identified with intuition as this term is used by Bergson.[1] If we take some of the terms which Bergson employs to characterize intuition, we may perhaps see the likeness and unlikeness. As for terms of agreement, direct awareness is non-symbolic as opposed to symbolic, acquaintive as opposed to descriptive, direct as opposed to indirect. But on the other hand, it is not absolute as opposed to relative; and, although it is not an awareness of an occurrent as related it may be an awareness of a relation. It does not grasp the occurrent, however, in any absolute sense as opposed to a symbolic awareness which views the occurrent as a totality of relations to other occurrents. It is not infinite as opposed to finite, nor perfect as opposed to imperfect. Though it is unanalytic as opposed to analytic, all awareness is unanalytic; thus it is not differentiated from any other kind of awareness. It is certainly not true as opposed to illusory. Truth applies only to symbolic awareness, as does error also; hence the direct awareness of an occurrent lies outside of the category of truth and error. So also with the characterizations of safe as opposed to dangerous, impractical as opposed to practical; these do not properly apply to direct awareness, for such awareness must always be supplemented by the symbolic awareness. In fact, the direct awareness cannot be properly characterized as knowledge at all; hence all of the attributes of knowledge are strictly inapplicable. It certainly does not give us a mobile reality as opposed to a fixed reality, nor a qualitative realm as opposed to a quantitative. To recognize reality as qualitative or as quantitative, or as fixed or as moving, always involves symbolic awareness. Mobility, fixity, quality, quantity—all of these are characterizations of occurrents and imply symbolic interpretations.

Thought

In the actual situation under consideration intuition is associated with *thought*, and we must now turn to a consideration of this aspect. But thought is the awareness of *symbols* and we cannot determine its nature until we have analyzed this type of

[1] *Introduction to Metaphysics* (New York, 1912), *passim.*

entity. Since the analysis of symbols is not to be given until chapter VII, our discussion must necessarily be brief.

When an occurrent has the property of *meaning* and when we are aware of the occurrent as having this property, then we are said to be thinking about the occurrent. Since an occurrent which has the meaning property is a symbol, we may avoid ambiguities by saying that we think about symbols. But one should not allow this verbal substitution to obscure the complexity of the situation. Every symbol *is* an occurrent, and as such has a definite character. For example, the symbol " man " has the character of being written " m-a-n " and of being sounded in a particular way. Hence in order to recognize the symbol as " man " rather than " dog " or " table," one must characterize it. But this process of characterization is precisely the activity of deciding upon a symbol which is applicable to it. In this way the act of identifying a symbol (apart entirely from the *meaning* of the symbol) can be shown to involve the three sub-aspects of knowledge, one of which we are now considering. Specifically, in order to recognize the symbol " man " for what it is, one must be aware of the occurrent, man, of the symbol by means of which this occurrent may be identified, and of the applicability of this symbol to the occurrent. Thus thought involves intuition, thought, and the awareness of applicability—all in the mere identification of the occurrent to which the meaning is attached. Before we can know the meaning, we must know the character of that with which the meaning is associated; but to know this character we must think about a symbol which portrays the character.

But thought is even more complex than this. For after we have determined the character of the occurrent to which the meaning is attached and have determined the fact of the attachment of the meaning, then we have the necessary grounds for the recognition of the " realm " in which the occurrent lies. More precisely, we know that the occurrent is a symbol and thus exhibits all of the general properties of symbols. This is the same general kind of recognition as the awareness that the occurrent is real or unreal, present or past, abstract or concrete.

This peculiar epitomizing character of thought would seem to argue for an inherent infinite regress in the determination of the character of any symbol. Such a regress is certainly potentially present in every symbolic situation. Actually it is not troublesome, because the more abstract awarenesses do not enter consciously into the majority of situations. Usually we are not even consciously aware of the character of the symbol; we tend to accept it implicitly. In certain types of thought,— e.g., thinking in terms of images or pictures—the character of the occurrential foundation of the meaning is very important, for it mirrors the character of the referent. But in word symbols, though the character of the word determines the meaning, the former is not usually attended to in a conscious manner.

The recognition that an occurrent possesses meaning and is therefore a symbol, is very important. Although a symbolic occurrent is an occurrent and therefore possesses the form and content of an occurrent, it is meaningful and therefore possesses the form and content of a symbol. There is no contradiction here, for the occurrent possesses these forms and contents in different senses. As occurrent it possesses the form and content of an occurrent, but as symbol it possesses the form and content of the meaning. It has been the purpose of the preceding discussion to show that a recognition of the form and content of a meaning presupposes the conscious or unconscious recognition of form and content of an occurrent to which the meaning is attached.

Awareness of Applicability

The third aspect of observation is the awareness of the symbol for the applicability of the word " table " to the occurrent given. This is clearly a distinct type of awareness which cannot be reduced to either of the other two, though it is dependent upon them, since it could not possibly occur in isolation. It is the awareness which is involved in all verification and is, in fact, that awareness which is the foundation for all truth judgments. A symbol which is properly applied is true. But, unfortunately, this awareness is also merely the awareness of a symbol, and the question immediately arises as to whether *it* is properly applied. This is determined, we have seen, by the application of the tests of truth, which constitute a part of the most complete cognitive situations. We have excluded this feature of knowledge from present considerations on the ground that we seldom have instances of such highly adequate knowledge. Actually, one often tests the symbol for applicability without realizing that he is doing so. There are many situations, presumably basic in character, in which, given an occurrent and a meaning, one may be aware directly and with a high degree of certainty that the meaning does or does not apply to the occurrent. (More accurately we should say that one may be aware of the applicability or inapplicability, and of the symbol for this relationship, and of the fact that the symbol for the relationship does actually apply to the relationship. The infinite regress which is involved here is apparent). Probably the simplest illustrative situation is one in which a thought, i.e. an imagined, red is compared with an observed red; here one is given a symbol and an occurrent to which it is presumed to apply in the manner common to all images,—i.e., by resemblance, and he makes an immediate judgment, partaking of a high degree of certainty, that the thought red does correspond in all relevant features to the observed red. Very much the same sort of awareness must occur in the verification of certain word symbols, where the occurrent referred to is one of the so-called simples of experience, e.g., a flash of color, a simple shape, a tone, or an odor.

It is not necessary to knowledge that every situation should contain such a direct and certain verificatory phase. Error arises precisely because the verification of a symbol is usually an extremely complicated process. Hence, one sees the applicability of the symbol only in a confused and partial way. The readiness of the verification is determined by a large number of factors, some of which have to do with the general character of the symbols involved, and others of which have to do with the technical difficulties encountered in producing the occurrents for awareness. The former of these cannot be adequately considered until we have made an analysis of meaning. We may merely mention, therefore, such factors as the kind of symbol involved,—i.e., the way in which it refers to occurrents; the abstractness or concreteness of the symbol, which necessitates an indirect verificatory route through less abstract and less concrete occurrents; the extensiveness or minuteness of the occurrent symbolized, which necessitates an indirect verificatory route through more clearly observed parts and wholes; the scope of the symbol itself, i.e., whether the symbol is considered essentially in its unique content or whether it is considered as encompassing all or a large part of its structural relations; and so on. In the latter class are to be found the difficulties associated with laboratory techniques,—e.g., the production of the desired occurrents in awareness through the creation of the proper causes and conditions, the employment of instruments, etc. A discussion of these factors belongs in a laboratory manual rather than in a book on the logical structure of science.

Hence, it is in association with this aspect of awareness that doubt and error arise. Doubt arises when the symbol, by virtue of one or more of the features enumerated above, cannot be readily verified. Error arises, when in our haste to arrive at certainty, we assert the applicability of the symbol upon insufficient grounds. The insufficiency of the grounds may lie either in the limitations in our awareness of the symbol or in the incompleteness or vagueness in our awareness of the occurrent. Every symbol is an element of a complete symbolic system, and the content of any symbol is the totality of its relations to other elements in the system; hence it is not surprising that our awareness of symbols is usually of a limited sort. Likewise, every occurrent is an element of a complete occurrential system, and the content of any occurrent is the totality of its relations to other occurrents in the system; hence it is not surprising that our awareness of occurrents is usually of a limited sort. This seems to be the foundation for the claim of the absolute idealist that truth is a property only of the totality of symbols. But this does not mean, as many idealists assert, that every partial judgment is false. It means, rather, that every partial judgment (except, perhaps, those relating to the simples of experience) runs the risk of error, the risk being greater in proportion as the judgment is narrow and loosely integrated with other judgments.

One is tempted to summarize this discussion of the first aspect of observation by showing that the total situation can be neatly pictured in a symbol. Since observation is here described as the situation in which we identify a table as a table, one is led to believe that the elements of the total situation can be symbolized in their proper relationships through the proposition, " This is a table." The proper name then indicates the bare occurrent of which one is intuitively aware; the concept indicates the symbol which one thinks; and the verb indicates the symbol for the applicability of the concept to the bare occurrent. There is much to be said for this representation, since it does reveal quite clearly the elements of the cognitive situation and their proper inter-relations. But it must not be considered as an *exhaustive* representation. For the proposition " This is red " is itself a symbol,—viz., a propositional symbol consisting of more elemental symbols. Since the proposition is asserted, it has an implied applicability to the realm of occurrents. Hence, it is properly expanded into " ' This is red ' is applicable," or " This is ' this is red '." But this again is a proposition which is asserted to be applicable. Hence it follows that there can be no symbolic expression of the complete character of observation, for the judgment that the total content is correctly applied to the situation cannot itself be a part of the content. The symbolic character of knowledge must therefore be " seen " but not talked about.

It has been the aim of the discussion in this chapter to show the nature of awareness and the extent to which it is under control. If awareness is under control, then knowledge, which is the awareness of a certain kind of content, is also under control. But although awareness is partially under control as to intensity, kind, and locus, there are distinct limitations to this control. These limitations are to be found in the character of the realm of clearly given occurrents, which constitutes the ultimate reference of awareness and therefore the ultimate foundation of knowledge. Occurrents cannot be made any more clear than they are actually given in their clearest form in nature; they cannot manifest themselves in realms other than those given by nature; they can permit of selection only within the realm in which they are given. Thus the ultimate starting-point and the final resting point of all knowledge must be the realm of clearly given occurrents.

But nature also reveals herself in the realm of obscurely given occurrents, and it is part of the task of science to determine the character of this realm as well. This realm, however, is essentially antagonistic to knowledge. Its obscurity is inherent and not merely a matter of our awareness. Hence, we must endeavor to find some way of overcoming this irrationality. This we do through symbols, which are those elements of the obscure realm which have an outstanding referential quality. We then substitute for the obscure occurrents symbols, and we define the symbols through their reference to the realm of clearly given occurrents. This gives us a way of becoming aware of the obscure occurrents through symbols which are relatively precise. We have thus inserted a factor of control into awareness and are able to call up at will a symbolic content when the occurrential content is unyielding. This becomes a means through which we subject knowledge to control and are enabled to replace inadequate knowledge by that which is more adequate. But although the selection of symbols is arbitrary, their definition is not; in every case there must be designated an operational route through which they may be shown to have locus within the realm of the clearly given. This introduces the notion of operations, which is properly the topic of the next chapter.

CHAPTER VI
OPERATIONS

Our approach thus far has been structural. We have attempted to analyze knowledge into its elements and their relationships, and we have found it to consist of certain kinds of awareness and contents in a peculiar juxtaposition. But we must now forsake this method for one which is less abstract. We must recognize that knowledge is not the eternal contemplation of the realm of occurrents. On the contrary, it is a *series* of such contemplations. Knowledge is indissolubly united with knowing. Reflection is an activity which goes on in time; it is not static but dynamic. The solving of a problem is a process in which the investigator is aware of a succession of elements, and aware of these elements in a succession of different ways. Furthermore, he is often aware of the succession as distinct from the elements; thus the movements of transition from one content to another may also enter into awareness. One may be aware of contents as developing. In this chapter we shall attempt to make clear this dynamic character of thinking.

A convenient notion for characterizing this movement of reflection is the concept of *operation*. An operation may be defined as an act by which awareness passes from one content to another. Thus an operation is a change in the locus of awareness; it is a shift of attention. But since a change in the content may involve a change in the character of the awareness, an operation may be a change in kind of awareness as well. If we consider the total realm of occurrents as the realm of potential awareness, then an operation is a selective act by which the locus of awareness is changed. If we use the figure of a searchlight playing upon objects, the movements of the searchlight from one object to another correspond to operations. The total realm is given, but not given in awareness. The succession of contents-in-awareness and of transitions constitutes the temporal feature of thought.

The structure of operations is therefore that of relational-complexes. Presumably the best analysis of this structure would be one which reveals it as a triadic complex consisting of that which is operated upon, the operator, and the resultant of the operation—all united in such a way that when the operation is performed upon that which is to be operated upon, the outcome is the resultant of the operation. A simple example may be drawn from mathematics. Given 3, we may perform the operation of addition by means of the operator 2 and obtain 5 ; the relational complex is then triadic,—viz., 3 plus 2 equals 5. But, as a matter of fact, in a large number of operations it is impossible to designate the operator as distinct from the operation ; in all such cases the relational-complex is more accurately analyzed as a dyadic complex. For example, when we generalize we perform an operation of generalization upon something which is specific and obtain something which is generic. In the same way unclassified entities may be classified to produce a class, unordered elements arranged to produce a series, unmeasured entities measured to produce a ratio, propositions inferred from other propositions to produce implicates. Hence, it seems advisable to consider the general structure of operational-complexes as dyadic, with the understanding that triadic complexes may be reduced to dyadic by including the operator within the operation and thus specifying a more complex operation. Then if a more detailed analysis is desired, the complex operation may be re-analyzed into the operation and the operator. An operation may then be defined as *the temporally successive awarenesses of the elements of a dyadic relational-complex.*

Kinds of Operations

There then arises a very important distinction. Operations may be *conscious* or *unconscious*. Roughly speaking, operations are conscious when we are aware of the act of transition and unconscious when we are not. In the former case the contents of the successive awarenesses seem to be related, and it is this fact of relation which accounts for the succession of the contents in awareness. The content of the second awareness seems not merely *to follow* the content of the first awareness but *to follow from* it. We seem to know in advance of the appearance of the content of the second awareness that it is to appear and approximately what content it is to have. It seems to arise through *derivation*. But in the former case the contents of the successive awarenesses seem to be unrelated, and there is consequently no apparent reason for the succession of contents in awareness. The content of the second awareness seems to thrust itself into the mind without being in any way derived or extracted from the content of the first awareness.

Although the distinction as here formulated has a certain working value, it is not sufficiently precise. The important difference is not the fact of the presence or absence of the awareness of the transition but whether this awareness does or does not precede the awareness of the second content. If the awareness of the act of transition does precede the awareness of the second content, then the latter seems to be derived and to arise in consciousness as a result of the act. But if the awareness of the act does not procede it, then it seems not to be derived but to arise in consciousness spontaneously. This may all be expressed in terms of relations rather than acts of transition by saying that in both unconscious and conscious operations we are aware of the *referent* before we are aware either of the *relation* or of the *relatum*. But in the case of conscious operations we are aware of the *relation* before we are aware of the *relatum*, and in the case of unconscious operations we are aware of the *relatum* before we are aware of the *relation*. This means, in the latter case, that we may not be aware of the *relation* at all, even after we are aware of the *relatum*. Hence, the problem is essentially one of whether the act of transition does or does not enter into awareness anterior to the awareness of the related element. If it does, the operation is conscious; if it does not, the operation is unconscious.

But there is another important distinction between kinds of operation which cuts across the above distinction. There are operations of *invention* and operations of *discovery*. This distinction does not suppose any awareness of the operation anterior or posterior to the awareness of the related element; hence there may be, as we shall see, unconscious inventions and discoveries and conscious inventions and discoveries. In fact, the feature which differentiates inventions from discoveries does not lie in the character of the operation itself, except indirectly. The distinctions between inventions and discoveries are three in number.

(1) Entities obtained through invention are relatively more obscure than the entities from which they were obtained, but entities obtained through discovery are usually as clear as the given entities. Thus the distinctive feature is the relative obscurity of inventions and the relative clarity of discoveries.

(2) Contents obtained through inventions have an inherent reference to occurrents which are more clearly given, while contents obtained through discoveries are self-contained. Thus the obscurity of invented contents is a specific kind of obscurity; it reveals a reference to the source to which one must go to remove the obscurity. The content of an invented entity is determined by the nature of something from which it has been derived; it is thus reducible to that other entity and the operation of derivation. Its nature becomes clarified through the designation of a more clearly given occurrent with which it has a specifiable relation. But when the operation is one of discovery, although we can see in the operation something of a directive character, we find in the derived content a character independent of the operation performed. Discoveries are often surprises; inventions are never such.

These two criteria are differentiations of the two types of operation upon the basis of the entities derived through the operation. But there is another criterion which is based upon a recognition of a feature of the operation. (3) If in the awareness of the operation we discern that it exhibits certain properties of objectivity and publicness, then, however obscure may be the related element, we tend to speak of the operation as one of discovery rather than of invention. Thus constructs which are universally and repeatedly derivable from given occurrents seem to lose both their obscurity and their referential character, and hence are considered to lie within the realm of the clearly given. What this means is that if the two former criteria of the distinction between inventions and discoveries fail of application, I resort to an examination of the operation itself. If I find that the operation can be repeated an indefinitely large number of times, and if I find that others can perform the operation and obtain approximately the same result, then I am led to conclude that what I am aware of is not merely the operation of derivation but the relation of connection between the given element and the resultant entity. It then appears to me that I am merely noting the relations of the given element, following out these relations, and discovering what occurs as the terminus of each relation. Since I can then perform the operation rapidly, I soon learn to perform it habitually; thus I forget that I perform the operation and I become convinced that I am discovering the entity.

We may now see how there may be four main kinds of operation,—viz., unconscious inventions, unconscious discoveries, conscious inventions, and conscious discoveries. The distinctions between them will not always be sharp, and we shall therefore employ illustrations as aids to understanding. The most clear-cut cases seem to lie within the classes of conscious inventions and unconscious discoveries. In fact, one easily gets the impression that all inventions are conscious and all discoveries are unconscious. Inventions seem to be devices, and hence to require a deliberative act of devising; but discoveries, because of the stubborn features which they exhibit, seem to be things which we stumble upon without advance knowledge either of their existence or of their features. It will therefore be necessary to find genuine cases of unconscious inventions and of conscious discoveries, if these are not to be relegated to the status of null classes.

Unconscious Inventions

Unconscious inventions will reveal the following features: The awareness of the given entity will be followed directly by the awareness of the resultant entity, with only an obscure and later awareness of the character of the operation of transition. Here the spontaneity of the occurrence of the secondary element seems predominant. But although the resultant entity occurs suddenly, it is not sharply defined; we are aware that something has entered into consciousness, but we are not quite sure just what it is. Furthermore, the element thus obscurely given seems to have an important reference to that from which it has been obtained; it thus suggests that its meaning is to be made more precise by a further examination of the situation out of which it arose and by an attempted ascertainment of the method of derivation. It suggests that it is to be operationally defined, and that its locus of definition is precisely the original occurrent from which it was operationally obtained.

The best example of unconscious invention can be found in the activity of the creative imagination. This should not be limited to the undisciplined imagination or day-dreaming of our unoccupied moments, but should include the activity of the creative artist as well. In the former type of rumination or mind-wandering the features of unconscious invention are clearly revealed. It is unconscious because we never know how it starts. Presumably it has its origin in something perceived or thought about. But we discover that the transition to an imaginative entity *has taken place*; we do not resolve to imagine and then set the proper activities into operation. But throughout the whole process we are aware both of the obscurity of the succeeding contents and of their artificiality. We do not often mistake reverie for perception because the former is inherently indistinct and lacking in vividness. Furthermore, we are more or less distinctly aware at all times of the fact that we have modified reality in some way, usually in the direction of idealization around our basic desires. Thus we see that the imaginative entities have a significant reference to reality, and have their contents determined thereby. In the case of the creative artist the situation is approximately the same. In a moment of exaltation the artist suddenly finds himself in possession of a guiding idea—obscure as to its outlines and obviously an idealization of something actual. Thus it exhibits the spontaneity of unconscious operations, and the vagueness and reference of inventions. Its content must then be made more precise by determining specifically in what ways it is a modification of the actual.

Another example of unconscious inventions may be found in illusions, if one eliminates those cases in which there is a peculiar vividness and persistence of the phenomena; the cases of this latter kind are more properly classed with hallucinations and belong in the category of unconscious discoveries. An illusion may appear with the same spontaneity as an observed object; but as we turn our attention to it, it becomes relatively obscure. Furthermore, we find that its content is explicable in terms of that of which it is an illusory counterpart plus the operation performed by the observer. When we know the object which created the illusion and the character of the total situation of which the observer with his spatial location, his specific sense organs, his past experience and training are a part, we can derive the content of the illusion. Thus the illusion is explicable in terms of a certain operational route. But since the complexity of the factors entering into the total situation is so great as to make a repetition of the operation and the result a practical impossibility, the illusory object tends to retain its fictional and constructional character; we do not feel obliged to find a place for it in the realm of real occurrents.

Conscious Inventions

Conscious inventions will differ from these in the following ways : The awareness of the given entity will be followed by a fairly clear awareness of the operation, all of which will occur in advance of any clear awareness of the resultant entity. The secondary element is then given more obscurely than both the given element and the operation. And it has this same peculiar referential character; it seems to be dependent upon the given element and the operation. But in this case we have deliberately performed a certain operation, knowing in advance that the character of the resultant entity is to be determined thereby. Thus we are impressed with the constructional character of this element; it seems to be our own creation, determined only by the character of the given and the operation of derivation.

The fact of a continuity between conscious inventions and conscious discoveries makes the selection of pure examples of either type of operation difficult. Let us attempt an abstract characterization of the latter before turning to illustrations.

Conscious Discoveries

Conscious discoveries will reveal both the direct awareness of the operation anterior to the awareness of the secondary content, and the clear and distinct awareness of the secondary entity as possessing an independent content. Conscious discoveries are explorations of the realm of the clearly given. This realm is a complex of distinct entities, having distinct relations. Awareness explores this realm, just as a traveler explores an island by following out the spatial relations and discovering what physiographical features are exhibited as he moves from place to place. If he is a good explorer, he keeps his bearings; i.e. he knows at any minute where he has come from and where he is going. Though he does not know at any moment what he is going to find, when he does make a discovery the features of the discovered element are then immediately and clearly given; he is not obliged to recall how he came, in order to note what he is seeing.

To find illustrations of conscious inventions and of conscious discoveries is to reflect an individual rather than a universal point of view. All obscure occurrents have the features of inventions; but the more obscure they are, the more certainly they lie in the class of inventions; and the more precise they are, the more reasonably they may be located in the class of discoveries.[1] This is due to a point of view to which we have already called attention, viz., when an invention is obtainable through an operation which is public and repeatable, the entity tends to pass from the obscure realm into the realm of the clearly given. Hence, in the discussion of this topic one is obliged to construct a series of such a character as to contain the entities which are quite certainly inventive at the extreme left and the entities which are quite certainly discovered at the extreme right, with all other entities arranged in between in their proper order. Then the indication of a cut in the series which would put all invented entities on the left and all discovered entities on the right would be quite arbitrary. It would certainly vary for different individuals who might draw it. And it would vary in

[1] Cf. H. Vaihinger, *Philosophy of 'As If'* (New York, 1925), p. 88.

time, since inventions tend to become discoveries as the operations are repeated. But, in spite of the tentative and speculative character of such an enterprise, I shall undertake to carry it out. It should be considered merely as suggestive of the solution.

In the construction of such a series an important principle must be recognized. Inventions are not supposed to make up a part of the real world, but discoveries are. Hence, we shall have no difficulty with the extremes of our series. Entities at the left will be inventions and will not exist; hence, we shall not trouble ourselves about how to locate them. Entities at the right will be discoveries and will exist; hence, we shall not have to trouble ourselves about them. But entities in between are not certainly non-existent, nor are they certainly existent. We cannot make them non-existent, for they might prove to exist; and we cannot make them existent, for they might prove not to exist. The principle which comes to our rescue at this point is the principle of symbolic reference. *If an occurrent is not known to exist, it may always be considered as the referent of a symbol without prejudice to its existential status.* Symbols are instruments through which we talk about occurrents independently of their existence or non-existence. Hence, the *safest* way in which to talk about the result of an operative act is to say that it is the referent of a symbol. Then, according to the principles of differentiation discussed above, the referent of the symbol will be an invention if its content is inherently obscure and relative to the situation out of which it arose, and if the operational route employed in its production is private and unique. Otherwise it will be a discovery.

Among the most certain of the conscious inventions are symbols themselves. However, the certainty of their location in this class of entities is dependent upon taking them from one point of view only. Symbols are creations of individuals. I can invent any symbol I please to represent any occurrent I please. The only limitation is that the symbol should represent the occurrent *for me*. Hence, in the invention of a symbol I find myself confronted with an occurrent for which there is no symbol; and I then pass to the awareness of a certain image, or noise, or group of letters to which I attach a meaning property.

Ever afterward I may use the symbol as a substitute for the occurrent. Here the operation is clearly one of conscious invention. It is conscious because I am aware of the given occurrent, of the need for a symbol, and of the general character of the act which must be performed in order to create a symbol. As a result of this awareness I find myself aware of a symbol. But the symbol is an invention, because it is essentially vague as to content and has a definite and unmistakable reference in the determination of its meaning to the occurrent from which it was derived.

But if one looks upon symbols as social entities and as instruments of communication, they cannot be placed so certainly in the class of inventions. If symbols were all images or icons,[1] this problem would not arise. Images are private and incommunicable. But words are not private. Hence, in devising and using word symbols I am under certain obligations. I must give them meanings according to approved techniques; I must retain the meanings of symbols which others have given to them; and I must employ symbols without violence to the recognized rules of syntax. But this means that the operational derivation of a symbol has become something which is public and repeatable. Anyone may derive a symbol from an occurrent, provided he is willing to abide by the rules of the game. But this is equivalent to giving symbols a kind of reality,—i.e., a social objectivity which is fixed by usage. In this way there arises some justification for speaking of symbols as discoveries rather than as inventions. The symbols become "attached" to the occurrents to which they refer, and are sometimes more easily discerned than the actual properties of the occurrents. We may know what an occurrent is called without knowing many of its properties. Since we discern the symbol by direct inspection of the occurrent, and define it as that which one normally becomes aware of in the presence of the occurrent, we unwittingly place the symbol within the realm of clearly given entities. Hence, we do not invent it but discover it.

Among the most certain of the conscious inventions are the so-called fictions and constructs of science,—e.g., perfect levers

[1] The definition of this notion will be given in Chapter VII.

and gases, ideal triangles and circles, isolated individuals, and utopias. In all such cases we are presented with an occurrent,—i.e., an actual state of affairs, in which we recognize certain inadequacies; and by a definite act, having a determinate character and a foreseen outcome, we pass to the awareness of a new entity in which these inadequacies, or as many of them as we please, are lacking. Is such an entity an invention or a discovery? In accordance with the distinction made above we may formulate the question: Is the referent of the symbol for such an entity non-existent or existent? The answer to the question is to be found in the content of the symbol. Clearly, it is both obscure and referential in content; i.e., we know very little about it and recognize that it is referential in character and definable in terms of that out of which it arose. Furthermore, the operational derivations are more or less arbitrary and hence neither public nor repeatable. Hence, the entity is an invention. But now if one were to make a persistent effort to define such an entity through a route which could be specified and characterized, and thus made public and repeatable, it would lose something of its obscurity and something of its remoteness from experience. For example, if we define mathematical points through Whitehead's method of extensive abstraction,[1] there is reason to believe that they are less obscure, "closer" to experience, and less arbitrarily defined than if we consider them merely as "positions without extent," or some such vague notion. More generally, if we define idealizations as abstractions from a series rather than as limits of series, there is more reason for including them within the real, since the fact that a series exhibits abstract properties is more clearly given than the fact that it has a limit. Perhaps the most characteristic movement in recent philosophy of science is the attempt to define, and thus to make public and repeatable, the operations involved in the creation of fictions. To the extent to which this is successful, fictions are passing out of the realm of inventions and into the realm of discoveries.

Probably the next items in the series should be the high abstractions and the high concretions. Do symbols for high

[1] *Concept of Nature*, chap. IV.

abstractions and symbols for high concretions have referents which are discovered or invented?

Here there is serious reason to doubt whether we are any longer concerned with inventions at all. To the novice high abstractions, if not high concretions, appear to be subjective. They seem not to be discovered in the real but projected into the real, either as forms of apprehension or as innate ideas demanding realization. But if one recognizes the strict continuity from the highest abstractions to the highest concretions, and the association of most occurrents both with more abstract occurrents and with more concrete occurrents, he soon becomes convinced that the operations of abstraction and concretion are not devices for producing constructions but routes leading to discoveries. It is generally recognized that there is a development in the ability to grasp abstractions. Early in one's training he must be continually referring to concretions in order that his abstractions may have "content;" but later he performs the operation of abstraction habitually and thus is able to locate the abstractions without noting the specific concretions. The same seems to be true of high concretions. Thus we may say that although high abstractions and high concretions arise first as conscious inventions they soon pass over into unconscious discoveries. This passage is dependent upon the fact that the operations are definable and thus may be made public and repeatable.

Intermediate in the series come the past and the future. Are the entities which are represented by symbols for the past and the future conscious inventions or conscious discoveries? They are not as clearly given as is the present, hence we cannot place them unequivocally in the realm of the discovered. Furthermore, they have a distinct referential character which necessitates our referring them to the present in order to determine their contents. Yet they are not purely arbitrary. In fact, there are definite operational techniques in terms of which, proceeding from the present, we may determine their contents. Furthermore, these operations are public and repeatable. They are based upon causal laws. Thus the future is definable as that which is obtained from a present occurrent when a causal inference is performed,— i.e., an inference from cause to effect. This indicates that the future is not merely that which comes after the present, but that which is causally determined by the present. So also the past is definable as that which is obtained from a present occurrent when a causal inference is performed,—i.e., an inference from effect to cause. This indicates that the past is not merely that which comes before the present, but that which causally determines the present. Then do we discover or invent the past and the future? The answer is arbitrary, since the operations are public and repeatable; when we all start with the same situation and perform the same operation, we usually get the same results. Thus we soon locate the past and the future in the realm of the relatively clearly given, and feel that we can explore them in much the same way that we can explore the present.

The remainder of the obscurely given entities may be presumed to lie somewhat farther along in the direction of discoveries. This means that they are relatively clearly, rather than relatively obscurely, given. It also means that their referential character is not so obvious; they seem to have features which they possess in themselves apart from their operational origin. It means, finally, that the operations through which they are derived are essentially public and repeatable; we can therefore become aware of the occurrents as often as we please and we can study them cooperatively. They thus seem to be given as definitely as the occurrents with which they are relationally united; hence they may be said to be discovered. However, they should probably be listed among the discoveries which are conscious ratner than unconscious, since there is still at this stage, presumably, a fairly clear awareness of the nature of the operation performed, which antedates the awareness of the entity derived. Here one would naturally place all occurrents in remote space, which can enter into direct awareness only by some process of communication. One would also probably locate in this class highly extended and highly minute occurrents. The operations involved in the derivation of highly extended occurrents are simply those of spatial or temporal synthesis—operations which are easily repeated and easily duplicated, and hence seem to reveal actual relational structures. Thus it is because space includes spaces, and time includes times, that we may perform synthetic operations upon small spaces and brief times to produce extended spaces and longer times. The same is true of the analytic operations in passing to highly minute occurrents. But in both cases there is still a relative obscurity associated with our awareness of these extremes. I am more clearly aware of the space of my table-top than I am of the space of the city of Chicago or of the space of a square millimeter of the table-top, and I am more clearly aware of the duration of the ringing of my telephone than I am of the duration of my life for an entire day or of a single sound of the bell included in the ringing. Thus, although these elements appear to arise in awareness with a definite content, they are still relatively obscure and must have their obscurity removed by the employment of operational techniques designed to direct the attention of the observer to occurrents which are given more clearly.

Last in the series of invention-discoveries will be those occurrents which are most clearly given of all—which are, in fact, as clearly given as are the occurrents from which the operation starts. Here the operation of discovery is simply an operation of exploration; the entities are "there" with a fixed and independent character, and we have only to bring them into awareness. The simplest example of this can be found in the examination of a field of vision. One starts with an element, and passes from this to another element. Here the spatial structure guides the operation, and it is the spatial character of one element of the scene that directs attention to another element of the scene, lying necessarily within the same spatial whole. Here the operation can hardly be distinguished from the relation; the shift of attention is simply the operational equivalent of the spatial relation. This is the characteristic of all operations which are public and repeatable.

Unconscious Discoveries

Unconscious discoveries will be discernible by virtue of the fact that the secondary element appears with the same spontaneity and suddenness as in the case of unconscious inventions. But in this case it not only appears without apparent connections with the given but it has a relatively clearly defined content. We are convinced, and sometimes startled, by the features which it exhibits. It has the same determinate form that the given, which we suppose ourselves to have left, also reveals. Thus we seem to have *come upon* another given, though we are not quite sure just how we managed to do it. Furthermore, the element which thus arises in awareness does not have the same referential character which it has in the case of inventions. We are less convinced that the element followed from the operation, and we see no particular reason to refer to the given element for the determination of its content. It is given clearly, and the problem seems to be one of the further determination of its content by exploring *it*, not by asking ourselves how we got it and how its meaning must therefore exhibit a certain dependence upon its origin.

The fact that conscious discoveries tend to become unconscious makes a sharp distinction between the two impossible. Of course, one should not say that a conscious discovery *becomes* an unconscious discovery, for a discovery is an act which occurs at a moment, and when it occurs it must be either conscious or unconscious; if it occurs at another time, it is another act and may then be of a different character. But if we try to classify occurrents and say that certain kinds are usually discovered consciously while others are usually discovered unconsciously, we attempt to generalize the operational methods, and we then come up against the fact that a certain kind of occurrent may sometimes be discovered consciously and sometimes unconsciously. Thus we must say that the realm of unconscious discoveries is identical with the realm of clearly given occurrents *plus* the realm of obscurely given occurrents when the operational techniques involved in their discovery have become habitual. Hence, we may stumble upon remote spatial objects, highly extended and highly minute occurrents, the past and the future, even high abstractions and symbols and fictions. This is simply to say that we often find ourselves thinking about these occurrents in much the same clear-cut way that we think about the more obvious occurrents; and since we are often not aware of the operational routes connecting them with the realm of the more obvious, we fail to see their dependence upon it. If someone claims to find these entities and to be able to think about them without referring them to certain more obvious occurrents upon which many can agree, we cannot object. But we shall probably not understand him when he *talks* about them, and we may therefore put it forward as a condition of communication that he show us how by starting with the occurrents which are clearly given to us (as well as to him), and by certain operational routes which are also clearly given (though possibly less clearly given) we may arrive at a situation in which we may be as clearly aware as he of the entities in question. Then, and then alone, will the entities be discoveries for us.

One of the most obvious cases of unconscious discoveries is the situation in which occurrents impose themselves upon our awareness by their intensity or novelty. They interrupt a chain of thought or reverie, and appear to intrude without relation to the content of consciousness at the time. Here there has been an operation, for awareness has changed content. But there was certainly no advance awareness of the transition, nor is there in most cases any posterior awareness of a relation between an element of consciousness and the intruding factor. In such cases we have the most perfect instances of discoveries ; if the intruding occurrent has any content at all, it must be found within the occurrent itself and not in its reference, for it has no reference except its spatial and temporal relations to other occurrents, and these were clearly not the basis for an operation which brought it into awareness.

Among unconscious discoveries one is also tempted to include acts of scientific discovery, so-called, by which theories and explanatory notions are brought into awareness. These are unquestionably cases of unconscious operations, for the long-sought ideas have a habit of popping into the mind on the most unexpected occasions, apparently without direction or control, and completely out of relation to what is in the mind of the investigator at the moment. But the question as to whether such operations are inventions or discoveries cannot be answered so conclusively. Often the theoretical notion occurs as a sudden illumination, as in the case of Poincaré's discovery of the fuchsian functions[1] or Darwin's discovery of the principle of natural selection.[2] Here the derived content seems to have the clarity which is necessary to permit us to characterize it as a discovery. But there is still the obstinate reference of the explanatory notion to a situation; thus, however clear the notion may be, it has a referential clarity. One is convinced that the notion arose ultimately out of a situation, since one must always engage in a long period of what Tyndall calls " preparing the imagination,"—i.e., collecting data. Furthermore, one must classify and arrange his data, and define and formulate his problem before the flash of insight occurs. Discovery never occurs to a disorganized mind. All of this suggests that although the psychological factors responsible for the occurrence of the notion in consciousness at one particular moment rather than another are still practically unknown, the theoretical notion has, nevertheless, important relations to the data.

It follows that much of what is called unconscious discovery can be shown to be unconscious invention. In fact, the verificatory techniques are precisely such as to establish this truth. For in verification we endeavor to show the relevance of the theory to the data. Since the theory is obscurely given, we indicate operational routes through which the obscurity may be removed; these routes direct us to situations where clearly given occurrents are located. Thus verification is simply a method by which we endeavor to bring into clear consciousness the operational techniques by which the theoretical notion might have been called into awareness at the outset. We insert the operation into the situation *post factum*. And it is characteristic of these retrospective analyses that they seem to affirm not merely how the sequence might have occurred, but how it actually would have been found to have occurred if one had attempted to discern the factors. All scientific theories seem so plausible after their discovery that one wonders why they were so late in appearing.

Later on[3] we shall make a distinction between the constructual

[1] *Foundations of Science*, pp. 387–8.
[2] F. Cramer, *Method of Darwin* (Chicago, 1896), chap. XV.
[3] Chapter IX. The question of unconscouis discoveries will be considered again at this point.

and the hypothetical methods, and only after this distinction has been introduced shall we be in a position to explain the sense in which most unconscious discoveries are reducible to conscious inventions. The two following situations seem to exhibit marked differences : Given a lever, one may by certain definite operations make it homogeneous, non-elastic, and weightless ; the result is an invention. But given heat, one may suddenly think of molecular motion. How does this awareness arise ? It is hard to say. The psychological factors at the foundation are still obscure. But it is clear that the notion of molecular motion is, at the moment, essentially void of clearly definable content. Of this alone we can be sure : if the notion of molecular motion is to explain heat, then it must contain at least such content as is demanded by what we already know about heat. Thus, knowing the character of heat, and the kind of explanation for which we are seeking, we can determine what must be the content of the notion of molecular motion. In this way both perfect levers and molecules arise out of essentially the same inventive technique. We shall call this the constructual method. The differences between the two notions arise when we pass on to increase the content of the notion of molecules through the hypothetical method; this operation is also inventive but of a different kind and not so intimately founded on the specific experimental situation at hand. Thus we *seem* to come upon something in the content of the notion of molecules which we did not put there, and this leads us to believe that we discovered it.

Normative Aspect of Operations

In all of this discussion there has been no mention of the normative aspect of operations. This omission has been justified, since it is important to know what operations are and that we do as a matter of fact perform operations before we attempt to show the right and wrong of thinking operationally. From the phenomenological point of view every operation is as good as every other ; since an operation is an act by which awareness changes content, any act which accomplishes this end must be an operation. This applies both to operations of discovery and to operations of invention.

But the referential character of the entities obtained through operations of invention places these operations upon a different level. Just so soon as they become public and repeatable, they take on a standard character. This is equivalent to saying that, given an occurrent and an operation performed upon that occurrent, there is then a right and a wrong way of attributing content to that occurrent. The right way will be based upon a recognition of the character of the relational-complex of which the operation is a reflection; the wrong way will be based upon a failure to recognize such a complex. The right way will produce such a result as can be corroborated by others performing a similar operation upon a similar occurrent, and by the inventor performing a similar operation at other times. The constancy in the outcome of the operation is an indication that the operation has been correctly performed. Relational-complexes must be internally consistent; i.e., given a term and a relationship of a certain character, the other term of the relationship is more or less definitely determined by virtue of its participation in this complex. Whether it is more definitely determined or less definitely determined will depend upon the determinateness of the relation. Hence, from the operational point of view, given a term and an operation, there will be a certain range of content which the resulting entity must possess, a certain range within which it is indeterminate, and a certain range which it cannot possess. If A is five feet long and B is longer than A, then B must be a long object, which has any length greater than five feet, and which cannot have any length less than five feet. If p is an abstraction from q, then p must have all of the implicans of q; it may or may not have certain other relations to q, but it cannot be incompatible with q. If M is a given whole and N is a part obtained by virtue of the inclusive relation of M to N, then the content which may be attributed to N is determined by the nature of M and the nature (or, as it is usually called, the *validity*) of analysis.

Hence, there must be expressible in certain cases a *rule* according to which one may judge whether an operation has been correctly performed. The rule is the expression of the structural relationship;[1] the act of performing the operation is a voluntary exposing of oneself to the necessity of the rule. An operative rule states that an entity of a certain kind, operated upon in a certain manner, produces another entity whose content can then be determined by the rule. The produced entity may be said to follow from the presented entity and the operation, when the operation has been correctly performed. The structural relationship and the rule are static; the act is dynamic. The relationship and the rule determine the necessity of the act. The rule directs the act upon the structural relationship, and the act is the manner of exemplifying the rule. The act is arbitrary in the sense that it may or may not be performed, but it is obligatory in the sense that, if it is performed, it must be performed in a certain way.

Our discussion thus far suggests that there will be at least the following main classes of rules for the performing of operations.
1. Rules for the formation of symbols in general.
2. Rules for the formation of symbols of what is remote in space.
3. Rules for the formation of symbols of what is past and what is future.
4. Rules for the formation of symbols of high abstractions and high concretions.
5. Rules for the formation of symbols of highly extensive occurrents (synthesis) and of highly minute occurrents (analysis).

Of these groups of rules the most important for our purpose is that concerned with the formation of symbols in general. All other rules will be based upon these. For, as we saw in chapter IV, the operational route by which we trace back an obscure occurrent to its locus always passes through symbols. To talk about the obscure occurrent we devise a symbol for it, and to talk about the operation we devise a symbol for it. Thus the rules for the determination of the remote, the past and future,

---

[1] Cf. Carnap's "*Übersetzungsregel*," *op. cit.*, p. 47. The dependence of the operation upon the structure of nature is obscurely recognized by Bavink (*The Natural Sciences*, pp. 228–35.).

abstractions, etc. will be the rules for the determination of symbols which mean these various types of obscure occurrents.

But we also saw that, since even the clearly given occurrents cannot always be conveniently produced in direct awareness, we devise symbols for these occurrents as well. There then arise two main types of problem : (1) How are the meanings of symbols determined by the occurrents which they aim to portray and by the operations of derivation ? This is the problem of the invention of symbols from clearly given occurrents. (2) How are the meanings of symbols for the remote, the past and future, abstractions, etc., determined by the more directly descriptive symbols from which they have been obtained operationally ? This is the problem of the derivation of symbols for obscurely given occurrents from symbols for clearly given occurrents. The fomer is an operation from occurrents to symbols; the latter is an operation from symbols to symbols. Neither of these operations is purely arbitrary; i.e., each of them may be rightly or wrongly performed. In each case what determines the validity of the operation is the accuracy with which the symbol for the operation reflects the actual relations of the elements concerned. This is determined by the extent to which the operation is public and repeatable. Thus the rules for the determination of meanings from occurrents will be ascertained by noting the ways in which people actually do build up such meanings ; the objectivity of the relational complex is social. So also the rules for the determination of symbols from symbols, although they appear to be of a different character, are really of the same kind. We must determine symbols for the remote, the past and future, abstractions, etc. by noting how people actually do attempt to give content to these notions in terms of the " here," the " now," moderate concretions, etc. Thus even in the latter case we are endeavoring to make our relation between the symbols reflect the actual relation between the elements concerned. Consequently the two types of problem can be shown to be only one general problem,—viz., how are symbols (i.e. simple symbols of simple occurrents and complex symbols of correlations and relational complexes of occurrents) derived from occurrents ? This is the problem of the empirical origin of meanings, to which we shall turn in the next chapter.

In summary it may now be seen that scientific thinking may be looked at either as a series of operational acts or as a complex operational act which may be analyzed into such a series. It is safe to say that the general problem of science is that of passing by an operation of invention from the realm of clearly given occurrents to the realm of symbols, whose contents are then determined by the occurrents and the operations of transition. Thus the total movement of science is one of creation; symbols are devised and given such contents as will make them retain their direct connection with the realm of the clearly given in spite of their capacity to portray also that which is obscurely given. Some symbols will retain that direct connection; others will be less directly related. It is the aim of science to increase the number of direct connections; thus where a symbol appears to have *no* connection with the realm of occurrents, or where it is only remotely or indirectly related, science must endeavor to ascertain through the operational method whether there is any more direct connection, and if so what that connection is. The realm of clearly given occurrents is the ultimate reference. To say that symbols are inventions does not mean that they are arbitrary and thus may contradict nature. Even an invention is a natural object; artifice does not oppose nature. Symbols are linguistic conveniences, human devices for grasping the character of the real; but the symbols are adequate only in so far as they reflect the character of the real. We find, therefore, that we are obliged to invent our symbols in a specific way and according to discoverable rules. We must turn immediately to the problem of the nature of symbols.

CHAPTER VII
MEANING

The problem of this chapter is at the same time one of the most basic and one of the most puzzling problems of philosophy. The extent of the literature in recent years devoted to its solution makes a thorough-going discussion of the problem in all of its aspects impossible within the scope of this book. I shall attempt, therefore, for the greater part, to avoid controversial material and to express, more or less dogmatically, the theory which seems to me to be the most adequate foundation for the construction of a logic of science.

It will be helpful to state in general terms the theory which I shall undertake to expound. Symbolic occurrents mean occurrents which are non-symbolic. But symbols are essentially devices constructed with a view to understanding the realm of non-symbols; they are thus fluid in character and dependent for their contents both upon those non-symbols whose natures they attempt to portray, and upon the operational activities involved in their construction. To show *what* a symbol means does not suffice; we must know also *how* it means. And to show *how* a symbol means does not suffice; we must know *what* it means in this manner. But both the *what* and the *how* of the meaning are contained in the symbol. Thus to reveal the meaning of a symbol one must show from what occurrent it has been derived, and by what operation it has been established. This may be called the theory of the *operational*[1] *origin of meaning*. We may proceed immediately to its more detailed exposition.

The triadic character of the meaning situation has often been pointed out. Where there is meaning there must be at least the symbol which means, the meaning itself, and the referent of the meaning. In any given situation the third of these elements, the referent, may not be found. But if the total meaning situation is to be genuine, there must be a referent. The triangular

[1] The word has been taken from P.W. Bridgman, *Logic of Modern Physics* (New York, 1927), p. 3 *et seq*. I believe, however, that I am using the word in a much more general sense than he is.

representation given by Ogden and Richards[1] portrays the way in which these elements are usually conceived to be related. It is clear that such representation demands not only three elements but also three types of relationship,—i.e., a relation which will connect each two of the elements. There must be the relation of the symbol to the meaning, of the meaning to the referent, and of the symbol to the referent. These, as the authors point out, are different kinds of relationship.

The essential inadequacy of this diagram, as I see it, lies in the supposition that meaning is a kind of element having relations to the symbol and referent. The situation then becomes complicated by the double necessity of characterizing these differing relationships, and of populating the world with the relationships and the meanings. It seems much simpler to suppose that meaning is merely a *way of referring* to an occurrent. Every symbolic situation, then, will contain a symbol, with its meaning or referential character, and a referent, which may not be given in the situation at the moment. These may be looked upon as three distinct kinds of elements, though they are very intimately interrelated. For an occurrent is symbolic only by virtue of the meaning property which it possesses, and a meaning property is never found apart from some occurrential locus. What makes an occurrent a symbol is precisely the fact of meaning. And although the occurrent which can be said to be the direct referent of the symbol need not be a part of the total symbolic situation at the moment, *some* occurrent which is at least the indirect referent of the symbol is presupposed in order that the situation as a whole may be meaningful. In other words, every symbol must have, directly or remotely, an occurrential foundation. Such a foundation may be either the situation from which the symbol was derived in the first place, or the situation to which one must ultimately go in order to determine the meaning of the symbol in the final reckoning. Thus there is a sense in which even the referent of the symbol is given in the total symbolic situation. In fact, it is the character of the referent which determines to a great extent the actual content of the

[1] C. K. Ogden and I. A. Richards, *The Meaning of Meaning* (3 ed.; London, 1930), p. 11.

meaning property; without an occurrential foundation a symbol would be without meaning. Hence, the interdependence of the elements in the symbolic situation can be clearly indicated by pointing out that an occurrent becomes a symbol by taking on a meaning character, which is in turn determined by the fact of a referent. But the meaning character must always be attached to an occurrent, and the occurrent which is meant is a referent of the meaning, precisely because of its participation in the symbolic situation.

From the psychological point of view the occurrence of a meaning is associated with a feeling of " vague anticipation : the mind is poised expectantly, awaiting something other than the thing, the symbol, which is immediately before it ; and this anticipation is *vague* because it is not accompanied by a belief that the object meant will appear or that it exists . . . . Though I cannot be said to turn my attention toward the thing I mean, since one cannot attend to something not presented to him, there is no doubt that I do more than attend to a symbol or an image. Indeed, I turn my attention away from the symbol or the image, and this constitutes the first step in preparation for the thing meant."[1]

We may now turn to a characterization of the meaning property. The simplest solution to the problem seems to lie in the supposition that an occurrent which means, as opposed to one which does not mean, possesses a unique property which may be called a *meaning property*. When the individual becomes aware of an occurrent as possessing this meaning property, he is *thinking* the occurrent ; and the occurrent itself plus the property is called a *symbol*. Thus thinking is a kind of awareness in which one becomes aware of an occurrent as occurrent and as symbol at the same time. When one becomes aware of the occurrent as occurrent, he is engaged in observation or imagination ; but when one becomes aware of the occurrent as possessing a meaning character, the type of awareness changes and he engages in thought. This unique meaning character has both *form* and *content*. The form is the general way in which the symbol means,—i.e., the way in which this symbol and a class

[1] Eaton, *Symbolism and Truth* (Harvard University, 1925), p. 23.

of similar occurrents mean occurrents; the content is the specific way in which this particular symbol refers to occurrents. For example, in the symbol " man " we have an instance of a general conceptual symbol; this symbol means in the same way as a large number of other symbols of the same type,—e.g., " woman " or " dog." The form of the meaning lies in the typical reference. But the symbol also means in a particular way not found in any other symbol of the same general character; viz., it means *man* and not *woman* or *dog*. The content of the meaning lies in the specific reference.

There are several advantages to this analysis of meaning. In the first place, it enables one to see how it is possible to speak of the content of that which is referred to, when it is not given at the moment. It explains how one can refer to occurrents *in advance*. The specific ways of referring are like the laws of projection; given the figure and the laws of projection, one can compute the projected figure. This is possible even though there is no intercepting plane and therefore no projected figure. So also if one knows the general type of symbol and one knows also the content of the symbol, he can determine in advance what the character of the occurrent must be. The laws of projection are like the laws of grammar; the figure to be projected is like the content of the meaning. Given a symbol with a content and a manner of referring, one can then compute the referent; the conclusion is prefigured in the computation—though it is not given in the computation, for the answer may be wrong, or it may not be obtained at all. The referent is given potentially in the symbol; the form of the symbol tells how to get it, and the content of the symbol tells what to get.[1]

Another distinct advantage of this conception of meaning lies in its application as a principle of parsimony. For not only the content but also the form of the meaning is located in the reference rather than in the referent. According to the traditional point of view (e.g. Russell),[2] the fact of different ways of meaning necessitates the supposition of different kinds of referents, —e.g., particulars, universals, and facts. But if the

[1] For some of these formulations I am indebted to Wittgenstein.
[2] As best expressed in his " Philosophy of Logical Atomism," *Monist* (1918-19).

solution of the problem lies in the fact that there is not one way of referring to many different kinds of things, but many ways of referring to only one kind of thing, these assumptions are unnecessary. All symbols, then, refer to occurrents, and that is all they can refer to. But different general kinds of symbols refer to occurrents in different ways; there are, for example, names, images, concepts, and propositions, all of which refer to the realm of occurrents but in different ways peculiar to themselves. And within their own specific ways of referring there are particular contents referred to, but *what* these contents are —though not *that* they are—is also a part of the manner of referring.

In the third place, it makes possible a general logic of symbols. The minimum requirement of a symbolic logic is that there should be types of elements and contentually different instances of the types, between which structures may be established. Symbols are typically different; names, images, concepts, and propositions have different syntactical rules, and must be represented by different symbolic forms. But symbols are also contentually different, for within the common forms there are individual symbols differing from one another in what they mean. We may thus establish structural relations between the individual symbols of a type, and between symbols of one type and those of another.

The essential difficulty with the conception of meaning here presented lies in the necessity of considering a meaning as a relationship—and yet not as the usual type of relationship, because it often exists in the absence of one of its terms. The advantage of the relational interpretation lies in the vivid way in which it makes clear the location of the content of the meaning in the relation itself. A relation is a connective, but it is also a connective of a certain kind; i.e., it has a content, as does any non-relational occurrent. So a meaning is a reference, but it is also a reference of a certain kind; i.e., it has a content. The disadvantage of the relational interpretation lies in the problem of attaching the meaning to its referent when the referent is not given. It might be advisable thus to represent the meaning property of a symbol not as a relation having a definite referent, but as an arrow, attached to the symbol at the one end, but pointing indefinitely away from the symbol. It is probably not essential to the understanding of the meaning property that it should be representable concretely; it is clearly unique among metaphysical elements; hence, any representation is liable to misinterpretation.

Certain features of the meaning property, as it exists in complete meaning-situations where the referent is also given, may be here indicated. In the first place, it always attaches to an occurrent and always terminates ultimately in an occurrent. The former is ordinarily a word, written or spoken, though it may be a gesture, or an act of any kind, or it may be any other occurrent which has taken on symbolic or referential character, such as a statue, idol, weather-vane, flag, or arrow. And since the referent of a symbol is always something in the realm of occurrents the terminus of a meaning will also be an occurrent. Thus meaning is a relationship between an occurrent which is symbolic and an occurrent which is not symbolic. In the second place, the meaning relationship is asymmetrical. If A means B, B does not in general mean A. In the third place, the relation is not transitive, for an occurrent cannot both be symbolized and be a symbol in the same sense; if we say that " chat " is a symbol for " cat " we cannot say in the same sense that " cat " is a symbol for a cat. Hence, we cannot say *as a result* that " chat " is a symbol for a cat. In the fourth place, to say that an occurrent is a symbol changes its essential character. By virtue of the meaning relationship, it takes on the form and content of a symbol although it retains its form and content as occurrent. Hence, to ask about the form and the content of a symbol is to ask an ambiguous question. Does one mean its form as symbol or as mere occurrent? Does one mean its content as symbol or as mere occurrent? A symbol, therefore, is any occurrent which possesses the unique property of meaning. But the possession of this character changes the nature of the occurrent so that it takes on an additional form and content determined by the character of the meaning.

The aspect of this interpretation of symbols which it is particularly important to emphasize is the dependence of the meaning of symbols upon the referents and upon the operations involved in passing from the referents to the meaning character. The essential contention here is that the nature of the meaning is a *product* of the nature of the occurrent from which it has been derived and the nature of the operation by which it has been derived.[1] It is because occurrents are of certain kinds, and because we try to symbolize them in certain general ways, that we have types of symbols. The traditional point of view has emphasized the dependence of meaning upon that which is meant. But it has overlooked entirely, or emphasized insufficiently, the fact of the dependence of the meaning also upon the character of the operation involved in the construction of the symbol. A symbol tells us not only *what* is meant but *how* the referent is referred to. Both of these are to be found in the situation in which the symbol was first devised. But the devising of symbols is one species of that general type of operational activity discussed in the previous chapter. Hence, symbols are meaningful occurrents whose natures are determined from other occurrents which they presume to portray, through operational routes indicating the way in which these occurrents are to be represented.

It is convenient to distinguish at this point between the *intension* of a symbol and its *extension*. These notions have been traditionally applied only to concepts, but they clearly have a general application to all types of symbols. The extension of a symbol is its reference to occurrents. This reference is of a particular kind indicated by the content of the symbol and of a general kind indicated by the form of the symbol. The intension of a symbol is its reference to other symbols.[2] Every symbol is a more or less highly integrated element of a larger and more inclusive system of symbols, and possesses relations of a variety of kinds (determined by the character of the symbolic system) to

---

[1] Cf. Bridgman, *op. cit.*, p. 5; Eddington, *Nature of the Physical World*, p. 254; V. Lenzen, *Physical Theory* (New York, 1931), p. 46.
[2] Cf. the pragmatic definition of meaning, e.g., "*the meaning of anything whatsoever is identical with the set of expectations its presence arouses.*" (C. W. Morris, " Pragmatism and Metaphysics," *Philosophical Review*, Vol. XLIII, p. 557). Cf. also C. I. Lewis, *Mind and the World Order* (New York, 1929), p. 83 : " The nature of a concept as such is its internal (essential or definitive) relationships with other concepts."

the other symbols within that system. The intension of a symbol is a part of its content, for the character of a symbol is determined to a great extent by these relationships to other symbols. One may characterize the content of a symbol which determines its extension as its *unique content*, and the content which includes its relations to other symbols as the *structural content*. Then every symbol refers to occurrents in two ways, directly through extension as indicated by unique content, and indirectly through intension as indicated by structural content. (The intensional reference may be considered a reference to occurrents, since the symbols which are related to the symbol in question have their own extensions).

One may then suppose a simple correlation between the content of the occurrent and the content of the symbol, and between the general operational derivation and the form of the symbol. According to this conception, the content of the symbol would reveal the content of the occurrent,—i.e., it would exhibit the specific way in which the symbol refers to occurrents; and the form of the symbol would reveal the manner of derivation —i.e., it would exhibit the general way in which the symbol refers to occurrents. This is a helpful interpretation, provided it is not misunderstood. However, there is danger of confusing the general manner of derivation with the specific manner. All symbols of a given kind are alike as to form, because they all refer to occurrents in the way which is common to the type of symbol. But within this type there may be variations in specific referential character, and these differences should be placed in the content of the symbol rather than in its form. Take the symbol, " perfect lever." It is part of the *content* of this symbol that it has been derived by abstraction or idealization from another symbol, " actual lever," which in turn has the *form* common to all such symbols, since it was derived by generalization from occurrents. Hence, the symbol, " perfect lever," means in the same general way as does the symbol, " actual lever " (i.e. through generalization); but the symbols differ esentially in content. We may say that all symbols for obscure occurrents have their contents as well as their general forms determined by operations. It is, of course, only the structural content which is so determined. From this it follows, as we shall see later, that symbols for obscure occurrents do not seem to have any unique content at all.

But another aspect of the meaning property of symbols must be pointed out, though it cannot be discussed in great detail until further considerations pertaining to the derivation of symbols have been introduced. The feature to which I wish to call attention is the *existential claim* which is contained in every symbol. If a symbol is a device constructed for the purpose of thinking about the realm of occurrents, then there must be contained in every symbol, in addition to the form which reveals the manner of referring and the content which reveals what is being referred to, a variable claim that that which is being referred to does actually exist within the realm which the symbol purports to describe. It is not only true that the symbol refers to a realm, but it is also true that there really is such a realm— at least the symbol makes such a claim, and this claim is clearly a part of the total meaning of the symbol. Such a claim must be variable, because it is determined by the operational origin of the symbol, and there is a great variety of operational routes by which a symbol may be established. All of these operations involve some loss in the forcefulness of the claim as to the existence of the referent. Most of our symbols are derived, not through direct examination of occurrents, but through operations upon other symbols, themselves presumably derived directly. These operations differ in their capacity to retain the claim of the original symbol as to the existence of its referent. Implication, for example, involves very little loss in existential claim, since, if a concept is exemplified, any implied concept will also be exemplified. But if the operation is based upon mere compatibility, or, worse, upon negational operations, such as incompatibility and contradiction, there is a more or less complete weakening of the claim, so that in the end we are quite unable to insist upon the actuality of the referent. The resulting symbol is, of course, still meaningful and still means in the same general way as the symbol from which it was derived; the weakening of the existential claim does not involve any loss of meaning. In fact, should the claim prove to be completely unjustified—i.e., should the occurrent referred to prove not to exist—, the symbol would still be meaningful. For there is a definite route by means of which it has been derived from something which does exist. Let us henceforth speak of this aspect of meaning as the *existential claim* of the symbol.

The discussion of meaning may be summarized under the following points : Meaning is a unique property attached to occurrents by virtue of which they become instruments of reference to occurrents outside of themselves. There are different, general kinds of meanings (forms of meanings) and different, special kinds of meanings (contents of meanings). The fact of the existence of meaning does not imply the existence of the referent of that meaning, though every meaning lays down an existence claim which is more or less forceful and which may or may not be corroborated. Meanings are determined by operations performed upon their referents, and are a function of the contents of the referents and the character of the operaton performed ; if the symbols are essentially descriptive—i.e., derivable directly from occurrents—, the form of the symbol is determined by the operation, and the content of the symbol by the character of the referent ; if the symbols are essentially suppositional or hypothetical—i.e., derivable indirectly from occurrents—, the content may also be determined by the operation. The *extension* of a symbol is its manner of referring directly to occurrents, and its *intension* is its manner of referring indirectly through its structural relations to other symbols. The *unique content* is that which reveals the unique content of the occurrent, and the *structural content* is that which reveals the structural content of the occurrent.

The most important types of symbols are images, or *icons* ; pointing symbols, or *indices* ; descriptive, or *general characterizing symbols* ; and structural, or *correlational symbols*. We may discuss briefly the general properties of each.

MEANING 149

Icons

Icons[1] or images are symbols whose meaning character is that of resemblance. Included in this type are all diagrams, metaphors, maps, photographs, memory images, projections, etc. The essential feature of such symbols is the similarity existing between the occurrent which is the symbol and the occurrent which is symbolized. By examining the character of the symbol as occurrent, we can determine the character of the occurrent symbolized. Here the " what " of the meaning is the actual similarity of color, shape, structure, function or quantity ; the " how " of the meaning is the general fact of resemblance. The operational character of the meaning is seen in the manner of its derivation. From a given occurrent by an act of duplicative construction an entity is created bearing a content determined by the given occurrent and the constructive operation. Thus the meaning content of the icon is determined by the occurrent and the operation. Had the occurrent been different, or had the operation been different, the meaning would have been different. The reference of the symbol is determinate and is based wholly upon the original formation of the symbol. The advantage of such symbols lies in their vividness, and also in their substitutional character ; their disadvantage lies in their inability to portray obscure occurrents.[2]

Indices

Indices or pointing symbols have no significant resemblance to their referents, and one cannot from the character of the symbol determine in advance the character of the occurrent which is meant. They are symbols which mean occurrents by direction of the attention to that which is meant. This direction of attention takes place, as Peirce[3] points out, by blind compulsion. Names ; pointing gestures ; arrows ; tags and labels ; such words as " this," " that," "these," " those," are all indices.

[1] I have taken the term from Peirce, *op. cit.*, Vol. II, par. 247. It will be clear in what follows that I have been much influenced by Peirce's analysis of symbols.
[2] The advantages and disadvantages of this type of symbol will be considered in greater detail in chapter XI. Cf. also Eaton, *Symbolism and Truth*, pp. 10–13.
[3] *op. cit.*, par. 306 ; cf. also Carnap. *op. cit.*, p. 16.

These symbols differ from all other symbols in that they have no content in the absence of the referent. All other symbols have both form and content in the absence of the referent; the form is the way of referring to the referent, and the content is that which points to the specific content of the occurrent. In all other symbols one can see the difference, though one cannot express it, between the meaning content of the symbol and the content of that which is symbolized. But names have as content only the content of that which is symbolized. The only symbol for the particularity of the occurrent is " this." But this symbol has no content independent of its referent. A proper name is the attempt to symbolize the fact of meaning without a symbol which possesses the meaning. " This " is merely the fact that a name is a name; i.e., it is the form of the meaning without the content. Obviously, it is impossible to employ such a symbol in the absence of its referent, for the existential claim is part of the manner of referring. In the absence of the referent there would be no way to determine it. Thus there is a special grammar for such words as " this." Their grammar involves the fact that their reference is unique in each situation of their employment, and that they are meaningless in the absence of the referent.

The form of the name indicates the operation involved in its derivation. It is an operation of particularization. This is an operation essentially of neglect, in which all relations of resemblance are denied and relations to the associated space and time are emphasized. For, although the particularity of the occurrent does not consist in its associated relation to space and time, these associations offer the most practical means for indicating that particularity. The particularity of an occurrent must not be considered to be an " aspect " of the occurrent in the same sense that a color is. Particuliarity is not a universal at all, but a way of referring to an occurrent. This accounts for the uniqueness in the application of a name; its content varies in each situation in which it is used. Yet it is not ambiguous; for whenever it is used, its referent is definite. It is clear that names are of no great utility in knowledge if one thinks of knowledge in its substitutional character. A name cannot be a substitute for an occurrent, for it can be used only when the occurrent is also given; and if the occurrent is also given, there is no occasion to use the name. Hence, the function of names is simply that of locating occurrents symbolized in other ways.

General Characterizing Symbols

Most important among symbols are general characterizing symbols. They may first be described by pointing out the ways in which they differ from icons and names. Since they have meaning in the absence of their referents, they are analagous to icons in an important way. But the specific nature of the meaning reference is quite different. Although the content of the meaning of the symbol is tied up with and determined by the content of the referent, the meaning relationship is essentially different. There is no apparent resemblance between a general characterizing symbol and its referent; i.e., a word or sentence which means an occurrent need not resemble that occurrent in any obvious way.[1] Thus one could not by a mere consideration of the content of the symbol, apart from its reference, determine what that reference is. But if one knew the nature of generalized characterization, and if he knew the specific meaning of a given generalized characterizing symbol, he could know in advance what the content of the referent would be. However, he would not determine this referent by looking about for something which is like the symbol. In this respect these symbols differ also from indices, for indices are not like their referents. But generalized characterizing symbols are not merely instruments of pointing, for they are meaningful in the absence of their referents. We can know what kind of occurrent to look for without at the moment finding any such occurrent. The symbol refers to occurrents not by blind compulsion but by characterization.

When we find ourselves compelled to say in a more positive way just what generalized characterization is, we experience certain difficulties. Both " generalization " and " characterization " seem to be essentially indefinable. Let us examine the former.

In harmony with our earlier distinction between form and

[1] This is not in disagreement with Wittgenstein (*op. cit.*, prop. 2.1). Symbols may be " pictures " of facts and yet not be icons.

content of a symbol, we shall say that the form is the general kind of way in which the symbol refers to occurrents, and the content is the specific way in which it refers to occurrents. Then we shall say that the form of such a symbol, when it is derived directly, is determined by the nature of the operation of derivation, and the content is determined by the content of the occurrent. But this means that the general kind of way through which such symbols refer to occurrents is itself determined through the operation of generalization. These symbols have the form of generalizations or universalizations. But what is the operation of generalization? It seems to be an indefinable operation involving essentially a passage from the occurrent in its existential and particular features to a symbol whose meaning then involves a reference back to the occurrent, and all similar occurrents, as mere potentialities and instances of the generalized character. The content of the occurrent is retained in the content of the meaning, but its individuality and its status as occupant of a given spatio-temporal situation are both lost. It is thus both an operation of neglect and an operation of addition. Although the symbol loses the individuality and particularity of the occurrent, it gains a generality which is not found in the occurrent. The occurrent is *this* occurrent and not *that*; the symbol is the meaningful reference to this occurrent or that occurrent, provided there is a similarity between the occurrents covered by the range of the generality of the symbol.

But what is characterization? Characterization is the retention in the content of the symbol of the content of the occurrent; this retention occurs only in such a way as is permitted by the nature of the operation; i.e., it is found in the symbol as a general character rather than as a particular instance. The content of the occurrent is not lost, for the symbol has a content; it is precisely the content of the meaning that determines the locus of its reference. But what is the content of this meaning? This expresses the essential difficulty in the notion of characterization. One cannot *say* what the content of any such symbol is, without employing another symbol of the same kind. In designating the referent of a name, one can point to a particular occurrent and say " this;" and if the occurrent does constitute the referent of the symbol, then it does so definitely and completely. So also if one employs an icon, although the referent need not be given when the symbol is used, the referent can be given ; and when it is given, it and it alone constitutes the referent. But in the case of generalized characterizing symbols one can only say that the referent is the whole of, or some selection from, the totality of things which are instances of the symbol in question. Thus the referent of the generalized symbol A is all A's, or any A, or some A's, or a certain A, etc. To specify the referent of the symbol we must determine the things which are A's and then specify the ones to which we are referring. But we can determine the things which are A's only by knowing what A means. Hence, the intension of A seems to be determined by its extension ; but to say what its extension is, we must know its intension.

Since generalization involves a loss in the particularity of reference, one easily gets the impression that generalized characterising symbols lack one of the prime requisites of symbols, —viz., definiteness of reference. They seem to portray the realm of occurrents inadequately ; for if the relation of the symbol to the situation is one-many, we cannot know through the use of the symbol alone which specific element of the situation is being referred to. A more adequate system would apparently be one in which there is a strict one-one correspondence,—i.e., one in which for every occurrent there is one symbol and for ever symbol one occurrent. But such a system as this, although it would gain in precision and adequacy, would lose in simplicity. A symbolic system which is adequate to all of the individual differences of the realm of occurrents would be as complex as that realm ; we should have an iconic rather than a characterizing system, and our ability to neglect systematically individual features of the world would be lost. But when we talk or think about the world we do not wish to re-create in thought all of its details ; we wish to grasp it schematically.

Attempts have been made to avoid this difficulty by supposing such metaphysical entities as general occurrents, universals, essences, classes, and the like. It seems unnecessary to populate the world with entities of this kind, since they fail to meet the problem which they are devised to solve. If one supposes universal or general occurrents, then the essential features of the realm of occurrents are lost, and a fruitful distinction between occurrents and symbols is lost. To introduce a realm of subsistence in which universals and classes may be located, is simply an attempt to retain the precision of the realm of occurrents by permitting all occurrents to be particular, and by introducing a realm of pseudo-occurrents which may be universal. Either of these suppositions can only postpone the difficulty. For then the two-fold problem arises : How are universals related to particulars, and how is it possible to symbolize universals ? Thus one must introduce a specific kind of symbol (viz. one which means a universal in its own peculiar way), and one must then show how it is possible for the entity thus referred to to enter into relation with another entity of an essentially different character. If one places the generality in the character of the reference rather than in the character of the referent, the problem becomes simplified, for then the manner in which the symbol means is precisely the manner in which the universal is related to the particular. Furthermore, the placing of the generality in the symbol seems to coincide with the fact that the meaning of the symbol is determined by its manner of derivation, and its manner of derivation seems to be a mental operation of generalizing. It is what we do to occurrents that enables us to refer to them in a general way ; we neglect their individuality and particularity in order to attend to the *kind* of content which they exhibit. But the *kind* of content is simply the meaningful way which a specific symbol has of referring to them, but not *merely* to them. Hence, the generality in the symbol resides in the fact of its being meaningfully referential toward occurrents not given at the moment but capable of being recognized, when they are given, by virtue of the content of the meaning reference.

Specific mention may be made of the attempt to locate the referent of a universal descriptive symbol in a *class*. This is clearly an attempt to show that the generality of the symbol lies not in the character of the meaning reference but in the character of that which is referred to, —i.e., to show that a universal symbol is one which refers in a way common to all symbols to an occurrent which is distinct from all other occurrents in being universal rather than particular. We saw in chapter IV that a class is a kind of complex occurrent which is formed by the dissociative union of two or more occurrents possessing similar unique contents. The fact of qualitative similarity with a correlated plurality seems to give us just the entity required for the referent of a universal descriptive symbol. But the inadequacy of the notion of class to serve as the referent of a general descriptive symbol can be readily seen. A class, as we have defined it, is extensional; i.e., it is an enumeration of elements. But an *extension* can be given only if the elements enumerated are given, and it is characteristic of general descriptive symbols that their referents are given not in actuality but only in potentiality. We can determine the class only if we know what *kinds* of elements are to be included. Nor can an *intension* constitute the referent of a general characterizing symbol, for an intensional class is not a class at all; it is simply the fact of meaningful generalized reference. Thus a class in extension is an occurrent, but a class in intension is a way of meaning. Hence, neither can constitute the referent of a generalized characterizing symbol, for the former contains only a part of what is referred to by the symbol, and the latter is not what is referred to but merely the manner of referring. Class as extensional is a metaphysical entity, but class as intensional is an epistemological entity. Though the former is the kind of thing which apparently could constitute the referent of the symbol—i.e., we could point to a class and say, " Here is the referent of the symbol "—, the fact of the generality of the symbol would be lost, for the class is potentially more than it is actually. And the latter is the kind of entity which could constitute the referent of the symbol only if symbols were always about ways of referring rather than things referred to—which is obviously not the case.

The simplest solution seems to lie in the supposition that general characterizing symbols are symbols which mean in a unique way. We may then say that the characterizing nature of the symbol lies essentially in its content while the universal nature lies essentially in its manner or form of meaning. A general characterizing symbol refers to a *kind* of occurrent, but a kind of occurrent is not another occurrent; a kind of occurrent is a way of referring to occurrents when their features are known in advance. One cannot therefore specify in advance what the referent of a symbol of this kind is to be.

Another way in which the nature of the meaning reference of generalized characterizing symbols may be explained, is in terms of the notion of *participation*.[1] Let us define participation as the relation of a symbol to a situation with which a referent of the symbol is associated. We shall recall that a situation is a block of space-time, and that such a block of space-time is usually occupied by (i.e. associated with) at least one occurrent. When we have such a state of affairs let us say that any symbol which means this occurrent participates in the situation. Let us then define *non-participation* as the relation of a symbol to a situation with which no referent of the symbol is associated. It follows that an occurrent whose character is symbolized may be recognized as absent from a situation. For example, if an occurrent is red, the symbol " red " participates in the situation of that occurrent, but the symbol " blue " does not so participate.

Then we may explain general characterizing symbols as follows : By the generality of the symbol is meant the loss in specific reference. Each symbol has its occurrential origin in one or more specific occurrents ; i.e., it is a generalization from one occurrent or from an enumerated group of occurrents similar as to unique content but different as to spatial or temporal location. The symbol refers back to the realm of occurrents, but not simply to the occurrent or occurrents from which it was derived ; it has become a potentiality of reference and thus refers to the total realm of occurrents,—for it refers by virtue of its origin to the occurrent or occurrents which suggested it in the first place, and it refers by virtue of its generality to all other occurrents. Thus, in the illustration, redness refers to all possible situations ; it refers to the here-now, in which red is found, and to the there-then, in which it is not found. If I specify r_1 and r_2 and r_3 and then indicate that the process is to go on without end, the referent of redness will be the totality of *all* occurrents. In other words, if I indicate merely the

[1] Cf. Whitehead's term " ingression," *Process and Reality* (New York, 1929), pp. 34, 63.

generality of the symbol and do not modify it by the fact of *characterization* I cannot specify properly what I consider to be the meaning of the symbol. But when I take into consideration also the fact of characterizaton, I can indicate that, *although the symbol refers to all situations, it does not refer to all of them in the same way.* There is a limitation to the universality, and it is precisely this limitation which is given in the fact of characterization. A situation may be characterized positively or it may be characterized negatively. The blue occurrent is characterized negatively by redness, and the red occurrent is characterized positively. Hence, redness refers to all occurrents which are red in a way in which it does not refer to all other situations; it must refer to reds positively and to non-reds negatively. This does not mean that there are occurrents which are non-reds possessing the same metaphysical status as occurrents which are reds. We have just seen that an occurrent whose character is symbolized may be recognized as absent from a situation. The occurrent which is non-red is simply the absence of redness in a given location. And this possibility of absence is the precise correlative of the generality in the reference of a symbol; it is because the symbol has a generality of reference that it is meaningful in the absence of an occurrent; and because it is meaningful in the absence of an occurrent, it possesses a type of reference to all situations in the realm of occurrents where such occurrents are absent. But the fact that the symbol is meaningful also in the presence of an occurrent of a certain kind and applicable to such an occurrent in a way in which it is not applicable to absent occurrents demands that the general fact of meaningful reference of characterizing symbols be differentiable into these two kinds of more specific reference. These different methods of referring are negatives of one another and are correlative in any general characterizing symbol; i.e., every such symbol will refer positively to the occurrents from which it was derived and to all situations in which similar occurrents are present, and it will refer negatively to all situations in which it does not participate.

We may make a few remarks about this distinction between kinds of reference. It is clear, in the first place, that negative reference is not to be identified with absence of reference;

i.e., when a symbol refers negatively, it does not thereby cease to be a symbol. Positive and negative reference occur as differentiations of a more abstract reference which is the type of reference of characterizing symbols. A symbol may possess both positive and negative reference at once, and thus be typically representative of occurrents. The loss of spatio-temporal specificity simply means a loss in the determinacy of the meaning reference; but the contentual feature of the symbol is not lost. This loss in the determinacy of reference implies that one can still say that the symbol means something in the realm of occurrents, but one cannot say where or when this something occurs. This is essentially the meaning of *possibility* as a character of generalized symbolic reference. A symbol is a possibility of occurrents because it inherently means in two distinct ways, and nothing in its character will enable one to determine which will be the reference to a given spatio-temporal situation. The awareness of the symbol must be supplemented by intuitive awareness.

In the second place the notion of negative reference applies only to descriptive or characterizing symbols. There are no negative names and there are no negative images. This is due to the fact that the negative reference is an implication of the generalized reference; it is because the symbol makes a claim to generality and thus presumes to refer to more than the specific situation which is its immediate referent that negation arises. Neither names nor icons make such a claim. The name points merely to that to which it points, and does not suggest a range of occurrents to which it might refer but does not. So also the image, while it may differ in content from many situations to which it would then seem to have a negative reference, does not claim any such reference; it is the picture only of its own referent to which it must then refer positively, or, more accurately, in a way which does not permit of differentiations into positive and negative.

Thirdly, negative reference may enter into direct awareness just as does positive reference; we may be aware of non-participation. This is the awareness of what an occurrent *is not*, and may or may not be accompanied by an awareness of what the occurrent *is*. In order to become aware of the absence of red from a situation I need not be aware of the presence of blue, or yellow, or something incompatible with red. But in another sense awareness of non-participation presupposes awareness of participation. This is demanded by the fact to which we have made repeated reference,—viz., that the meaning of a symbol must be found ultimately in a situation in which the referent is located. Thus to know that red does not participate in a situation implies that I know what red is, and this implies that I have observed red occurrents. In this sense non-participation can be recognized only if at least one situation involving participation has been experienced. This general fact of non-participation accounts for the fact that verification may be negative as well as positive.

In the fourth place, and as a result of the fact just mentioned, negative reference in a symbol determines a difference in existential claim. Those situations to which a symbol refers positively may be known to exist; there must be at least one such situation if the symbol is to be meaningful. But the existence of situations in which the given symbol does not participate cannot certainly be determined in advance of observation. The only evidence for the belief that there are such situations is the rather conjectural hypothesis that there are no occurrents which are everywhere; hence, it will always be possible to find a situation from which a specifiable occurrent is lacking. If the important property of high abstractions is their pervasiveness, then there is reason to believe that the negatives of the categories can never be illustrated. This may account for some of the difficulty which we experience in trying to think them.

Concepts and Propositions

The two most important kinds of generalized characterizing symbols are concepts and propositions. The difference between these two types lies in the manner of referring to occurrents. Though both are the same generic type of symbol (i.e. both refer through general characterization), they refer in different ways.[1] The simplest way of expressing this difference is to say

[1] Cf. Ritchie, *Scientific Method* pp. 33–35; Campbell, *What is Science?* pp. 40–49; also *Physics, the Elements*, pp. 39–47.

that concepts may refer to unanalyzed occurrents, whereas propositions must always refer to analyzed occurrents.[1] Concepts may also refer to analyzed occurrents, but if they do so they refer explicitly to one element of the structured whole and only implicitly to the structure itself, while propositions refer explicitly to the structure of the whole. Perhaps one may best express this by saying that concepts refer to occurrential elements of occurrential structure, while propositions refer to occurrential structures of occurrential elements. The ultimate referents are simple occurrents and occurrents related into structural patterns which in turn constitute occurrents of greater complexity. Simple occurrents can be referred to only by concepts, which do not contain in their symbolic form any specific indication of structure. But complex occurrents can be referred to either by concepts or by propositions. If they are referred to by the former, they are considered as unanalyzed wholes participating in larger complexes; but if they are referred to by the latter, they are considered as structures of simpler elements and the character of the structure is indicated in the meaning of the symbol. Thus if we assert " all men are mortal," there is clearly indicated a relationship between " men " and " mortal," and the character of the total symbol is essentially revealed in that feature; but if we consider " the mortality of men " or " mortals who are men," this relational character recedes into the background and the terms of the relational structure and its relational properties receive emphasis. This is true even where the concept is itself relational in character. Where we consider a relational concept,—e.g., " the fatherhood of X," there is less emphasis upon the relational character than there is if we assert " X is the father of Y;" in the relational concept the actual relatedness of the situation is lost and the consideration of a relational property takes its place. Relations determine attributes of the related terms, and it is these attributes which are essentially referred to when our symbol is conceptual, whereas it is the relation itself as an actively connecting entity which is essentially referred to when our symbol is propositional. This seems to be the foundation for the claim that propositions assert *facts*, while

[1] Wittgenstein, *op. cit.*, prop. **2.15**, *et seq.*

terms designate merely *elements of facts*. In line with our insistence that formal differences in symbols are fundamentally differences in the way in which symbols refer to occurrents rather than in the character of that which is referred to, we shall insist that there is no basic difference between that which is referred to by a concept and that which is referred to by a proposition, i.e., between an element of a fact and the fact itself. Both are occurrents, though of different degrees of complexity. Then we can say that propositions refer to occurrents in a way which clearly indicates the relational aspect of the complexity, whereas concepts refer to occurrents in a way which does not.

Another way of distinguishing between concepts and propositions is to say that concepts merely *refer* to occurrents, while propositions *refer* to occurrents and also *assert* them. I am inclined to think that this distinction is not nearly so basic as the one just mentioned, though it has a certain practical utility. It implies that the existence claim which is associated with every symbol is a part of the meaning of the symbol in the case of propositions but not a part of the meaning in the case of concepts. This seems to me to rest upon a more or less arbitrary employment of the term " meaning." We have seen that every symbol makes an existential claim, which may vary in forcefulness over a wide range, and we shall see later that the manner of deriving symbols through operations determines the forcefulness of the existence claim. Both concepts and propositions, therefore, make such claims. But it is clear that the existence claim in the case of propositions is more or less explicit; for every proposition, in assuming the assertive form, openly and explicitly reveals its existence claim. The truth of a proposition seems to be much more intimately connected with the meaning of the proposition than does the descriptive reference of the concept with its meaning. When one employs a proposition in discourse, he presumes, unless information is given to the contrary, that it means an occurrent, and that the occurrent exists. But when one employs a concept, he does so only by inserting it in a propositon. Thus the existence claim of the concept is lost in the more forceful claim of the proposition as a whole. Furthermore, one can make explicit the existence claim of a concept only by asserting a proposition about it—the proposition which states that the concept is or is not exemplified. Hence one must assert a proposition with an existence claim in order to show that the existence claim of a concept is not justified. Again, the symbolic relations between concepts are not so determinate with reference to the carrying over of the existence claim as are the symbolic relations between propositions. For example, given a concept A, known to have a justifiable existence claim, we can make no definite assertion as to the justifiability of the existence claim of non-A, obtained by contradicting A; such an assertion depends upon the abstractness of A. But given a proposition P, known to be true, we may conclude that the existence claim of non-P, obtained by contradicting P, cannot be justified. This follows from the fact that the negation of a concept results in a symbol having reference to all occurrents which are qualitatively other than those referred to by the original, but the negation of a proposition results in a symbol having reference to qualitatively different organizations of the same occurrents referred to by the original; if these qualitatively different organizations are incompatible, then the existence claim of P cannot be transmitted to non-P. For all of these reasons it seems advisable at least for practical purposes to emphasize the assertive feature of propositions and to neglect this feature of concepts. So long as this is considered merely a matter of emphasis and does not involve a denial of the existence claim of concepts, it is convenient for making clear the distinction between these two types of characterizing symbol.

Correlational Symbols

The final type of symbol is the correlational symbol. This is a symbol for the structure of the realm of occurrents. If the realm of occurrents is to be adequately portrayed in our symbolic system, there must be symbols not merely for the elements of the world but for its relations as well.

But correlational symbols are not merely symbols for relations. Correlational symbols must be distinguished from relational symbols. Both types of symbol refer to relations between occurrents; hence, both reveal the structure of nature. But there are two differences between them. In the first place, the former reveal the spatio-temporal structure of occurrents in a way in which the latter do not. If two occurrents are correlated, they have determinate spatial or temporal relations to one another. Hence, from knowledge of the location and nature of one occurrent and of the character of the correlation one could infer something of the location and nature of the other occurrent. Correlations are such as to guide awareness from one occurrent to another, and symbols for correlations are aids in the determination of the meanings of obscure symbols, for they direct the mind to the situation through which the obscurity may be removed. All associative and dissociative complexes are correlational complexes and exhibit relations of association and succession, distance and direction, spatial and temporal inclusion, and the like. Mere relational complexes are such as exhibit similarity and difference, quantitative variation, and the like; here there is no determinate location of the respective terms of the relations.

In the second place, correlational symbols determine relations between symbols while relational symbols do not. In other words correlational symbols determine the intension of symbols while relational symbols do not. Correlational symbols therefore have a double function: they represent the structure of occurrents in an indirect manner and the structure of symbols in a direct manner. Relational symbols represent the structure of occurrents in a direct manner and do not represent the structure of symbols at all. The indirect manner of referring which is characteristic of correlational symbols is indicated in the words which we employ to designate them,—e.g., implication, compatibility. These are *prima facie* relations between symbols, not between occurrents. Yet they do portray the structure of occurrents indirectly; for when one symbol implies another, the occurrents exemplifying the symbols have determinate relations to one another. Hence, one should say that occurrents are associated and the symbols implicationally correlated, but one should not say either that one occurrent implies another or that one symbol is associated with another (in this strict sense of the term " association ").

In this way correlational symbols reveal the structure of the realm of occurrents. But this structure is exhibited, as we have already seen, in the occurrence of associative and dissociative complexes. However, not all such complexes will be symbolized through this type of symbol. There are correlations which are relatively determinate and correlations which are relatively indeterminate. The determinateness or intimacy of a correlation is measured by the degree of definiteness with which the location and character of one element of the complex may be determined from a knowledge of the other element and the nature of the correlation. A highly determinate correlation is one which permits inference,—i.e., the attribution of content to the derived entity upon the basis of the given entity and the correlation. By inference is meant, not the carrying over of the same content to the derived entity, but any relatively high determination of its content through a knowledge of the rules guiding the operation. A determinate correlation will be one for which there are rules which must be obeyed if we are to give the derived entity a legitimate content.

That which determines the intimacy of a correlation is nearness in space and time, and repeated correlatedness. Two occurrents which are close in space are more likely to be determinately correlated than occurrents which are remote, and occurrents which are near in time are more likely to be determinately correlated than occurrents which are remote. So also two occurrents which are found repeatedly in such relations are more likely to be determinately correlated than two which are only once or only occasionally thus found. (This statement is not quite precise ; since occurrents cannot repeat, one ought to say that the symbols for the two occurrents are repeatedly exemplified in certain correlational complexes). These are expressions of our basic beliefs that there is no action at a distance, that causes which are distant in time must act through intermediate causes, and that what occurs many times is more significant in the world than what occurs only once.

We may then construct a table of complexes, passing from the most highly determinate to those which are relatively indeterminate. The precise order of terms here is not important. The purpose of such a table is to indicate the main types of complex to be found in nature. There are, of course, only two such types, the associative and the dissociative complexes. But the dissociative type is highly diversified as to kinds since there are many ways in which occurrents do not occupy the same situation. Furthermore since determinateness is dependent not merely upon closeness in space or time but on the frequency of the recurrence of the complex, each of the types of associative and dissociative complex will manifest three degrees of determination : (*a*) The correlation may be considered as occurring repeatedly in such a way as to permit us to designate each element as the sufficient and necessary condition of the other ; given one element we always find the other, and given the other we always find the one. (*b*) The correlation may be considered as occurring repeatedly in such a way as to permit us to designate one element as the sufficient condition only ; given one element we always find the other, but given the other we need not find the one. (*c*) The correlation may be considered as occurring only once or infrequently ; given one element we may or may not find the other, and conversely. This list represents a decrease in the order of determinateness.

1. Associative complexes. Here we have identity of space and time. In the most highly determinate form we have mutual transfer of contents. In the less highly determinate form we have the distinction between abstractions and concretions, which permits a more limited carrying over of content. In the least determinate form we have mere compatibility and practically no possible transfer of content.

2. Dissociative complexes involving, first, identity of space and brief time separation, or, second, identity of time and minute space separation. In the most highly determinate form the first type is cause-effect complexes in which the cause is the necessary and sufficient condition for the effect. Such complexes permit highly determinate inference to the past and to the future. Complexes of the second form are configurations and enable

us to infer to adjacent space from immediate space. The less highly determinate forms weaken the causal inference by virtue of the possibility of plurality of causes or of effects. Complexes revealing only occasional succession or occasional spatial adjacency are of little value as grounds of inference.

3. Dissociative complexes involving, on the one hand, identity of space with temporal inclusion, or, on the other, identity of time with spatial inclusion. These are complexes consisting of an occurrent in its spatial or temporal relation to its included or including occurrent. Here the validity of the inference will be dependent upon the extent of the including or included occurrent with reference to the given occurrent. If we pass from a moderately extended occurrent either to a highly or to a minutely extended occurrent, the inference is not highly determinate. The inferences in these cases are those formulated under the problem of the validity of synthesis and of analysis. In the highly determinate form (viz., part and whole always found together) relatively satisfactory inferences may be made; but where either one is sometimes lacking, or where they occur in this relation only occasionally, there are few significant inferences.

4. Dissociative complexes involving wider spatial separation with or without temporal identity, wider temporal separation with or without spatial identity, or both wide spatial and wide temporal separation. Here the inferential possibilities become practically nil.

A complete logic would contain a discussion of the meanings of symbols for all of the types of correlational complexes listed above. We should then know how to infer from an occurrent not only to its associates but to its abstractions and concretions, its spatial neighbors, its past and its future, the whole of which it is a part and the parts which it contains, and so on. We should then have an operational technique for determining the content of any obscurely given occurrent which makes any claim to reality.

Such a task is clearly beyond the range of the present book. When we turn, therefore, in the next chapter to a more detailed consideration of the empirical foundation of correlational symbols, we shall be obliged to make a selection. We shall confine our attention almost entirely to associative complexes, referring to dissociative complexes only as generalizations of specific dissociations. In other words, we shall say nothing about the special operational techniques involved in inferring from a given occurrent to what is elsewhere in space or time, or to what it includes or to what is included in it. These involve, in addition to the general principles of logic, specific principles relative to the character of time and of space, and to the character of complexes,—i.e., their heterogeneity or homogeneity, aggregational or integrative character, etc.

Unique among the correlating symbols is the relation of *participation*. There is some reason to believe that this is not only unique among correlating symbols but perhaps also unique among symbols in general. It does seem to belong within the category of characterizing symbols since it indicates structure. But it does not seem to possess the direct and indirect manner of reference which is the property of correlating symbols. It has only one referent, viz. the relation by which a symbol is attributed to a situation. It is thus a relation which spans the bridge between occurrents and symbols, yet it is not to be confused with the manner of referring which is the property of the symbols. For symbols have manner of referring even in the absence of a situation to which they may refer; but the relation of participation can never occur unless there is both a situation and a symbol. Furthermore, this symbol seems to be unique in that the very attempt to show what it means implies an understanding of it. For if the symbol called " participation " refers to the relation of a symbol to its situation, then if we are to understand what participation means we must find a situation which exemplifies it, and we must say that " participation participates in this situation." Thus we explain the meaning of the symbol by revealing a situation in which it participates; but this is clearly to employ a circular method of explanation.

These, then, are the main types of symbols. In each case the symbol is an occurrent having definite content and form as an occurrent. But it is also a symbol and thus takes on the form and content of a symbol. The content of the symbol is the specific way of referring and the form is the general way of referring to occurrents. The content of the symbol is the way of referring to the occurrent under consideration in such a way as to exclude other occurrents, though not necessarily *all* other occurrents; the form is the way of referring which the symbol possesses in common with a large class of other symbols. If I form an image of an occurrent I exclude other occurrents, but at the same time I refer to the occurrent in a way in which I might refer to many other occurrents. So also if I name an occurrent, or refer to it through concepts or propositions. In each case the general character of the symbol is determined by the operational route of its derivation and its operational foundation in the realm of occurrents. The operational foundation determines part of its content, and the operational route determines the rest of its content and its total form. Fixed operations determine fixed types of symbols, but varying occurrents determine a great variety of symbolic contents. All operations of symbolic derivation are inventive in character.

In the next chapter we shall continue this same general topic, attempting to show, not how isolated meanings are determined by isolated occurrents, but how the structure of the realm of symbols reflects the structure of the realm of occurrents.

CHAPTER VIII
MEANING : CORRELATIONAL SYMBOLS

Thus far in our discussion of meaning we have placed our main emphasis upon the general problem of symbolic reference. We have attempted to point out the main features of the meaning property, and have tried to show how the several types of symbol derive their meanings from occurrents by specific kinds of operation. In all of these considerations it has been important to emphasize the extensional character or the unique content of the symbols, for it is essentially the direct operations upon occurrents which determine the unique contents of the resulting symbols.

But all symbols, except indices, of which we shall speak immediately, have structural content, or intension, as well. The intension of a symbol is its relations, also symbolic in character, to other symbols, thus forming a complex or more inclusive symbol. For example, the intension of an icon is its relations, also iconic in character, to other icons which unite in this relationship to form a more inclusive icon. If I form an image of my living room, I find that the image content reaches out by means of relations which are themselves also part of the image to other images which represent the hall, bedroom, dining room, etc. By following out the structural relations of the image I can construct a more inclusive image representing a more inclusive occurrential situation. The intension of a concept is the totality of its structural relations to other concepts, e.g., what it implies and what implies it, what is compatible and incompatible with it, its relations to the concepts designating its wholes and its parts, its relations to the concepts designating its temporal antecedents and successors, etc. So also the intension of a proposition is the totality of its relations to other propositions which unite together to form a propositonal system. Thus there is given implicitly in any symbol, through its structural content, the totality of all symbols of the same kind. Some of these relations are highly determinate and thus affect significantly the contents of the respective symbols ; others are less highly determinate and consequently give us no essential information as to the contents of the related symbols. This intensional character of symbols is a very important feature of symbolic reference, and if we should stop our discussion at this point we should have told only half, and a relatively unimportant half, of the story.

The fact that every symbol has this intensional reference and thus endeavors to express not merely its "own" occurrent but the more inclusive situation of which the occurrent is a part, further complicates the problem of meaning. It means that the content of a symbol cannot be determined by a simple operation. When we have made our symbol adequate to the situation which it attempts to portray, we must keep in mind the fact that the task of science is not the construction of isolated symbols representative of isolated occurrential situations. The given situation is part of a larger situation, and the symbol which aims to portray it must be part of a larger symbolic system which aims to portray the larger occurrential situation. The representative character of any symbol is not simply its capacity to symbolize its own occurrent but this occurrent in all of its relations to other occurrents. Hence, a symbol is always derived with a view to its integration into a body of other symbols, some of them already established. This means that the content of a symbol is not fixed definitely by any simple act of derivation; it is rather tentative and precarious and subject to changes necessitated by modifications elsewhere in the symbolic system. Thus no symbol can be given definite structural content until its relations to other symbols can be determined, but none of its relations to other symbols can be determined until its unique content has been established. This necessitates a parallel development of the unique and structural contents of the symbols, each being subject to modifications in the other, and the total body aiming to represent the complexity of the realm of occurrents. It explains why symbolic systems are slow in developing, especially in those fields where the symbols are very complex in character and have many relations to other diverse symbolic elements.

The problem of this chapter is a continuation of the problem of the last chapter,—viz., that of determining the manner in which the content of symbols is derived from the content of occurrents through operational routes.

The question of unique contents has already been answered in the preceding chapter. The unique content of a symbol is its unique manner of referring to occurrents. The determination of unique content arises through a very direct operational route from the occurrent. In the case of names this route is called naming; in the case of icons it is duplication; in the case of concepts and propositions it is generalized description. All of these are attempts to preserve the content of the occurrent in the symbol and thus to make the symbols as adequate as possible to reality. Naming is the most direct of these routes, but it can be employed only in the presence of the occurrent and is of no value as a general symbolic form. Duplicating is an operational route which would seem to overcome these disadvantages, because the resulting symbol has meaning in the absence of the referent. But here generality is also lacking, and only certain kinds of occurrent may be thus referred to. In generalized characterization a symbol is devised which refers in a manner peculiar to its type : it refers to a *kind* of occurrent, which means that it refers universally through descriptive characterization.

It is with the question of structural contents that we shall be primarily concerned in this chapter. The structural contents of a symbol are its intension. Specifically we wish to determine the manner in which the intension of symbols is derivable from occurrents. The general fact upon which this is all based is that the structural relations of symbols portray the structural relations of occurrents. The unique content of the symbol reveals the unique content of the occurrent; the structural content of the symbol reveals the structural relations of occurrents ; and the form of the symbol designates the manner in which the symbol refers to the entire occurrential situation. Thus we must not determine the structural feature of symbols by examining isolated occurrents ; it is occurrents as uniting in associative and dissociative complexes that enable us to determine structural contents of the symbols. Here, as also in the case of our consideration of occurrents themselves, we do not need to mention *all* of the relations between occurrents; it is only the most determinate of these which are significant for the determinaton of symbolic structure. The more specific occurrential relations can be designated by relational symbols which then take on structural relations determined by the way in which the relational occurrents unite to form associative and dissociative complexes with the non-relational occurrents.

Even this problem was touched upon in the last chapter. The problem of the operational origin of the intension of symbols is precisely the same as that of the operational origin of correlational symbols. Correlational symbols reveal the fact of structural relations; they reveal directly the fact of structural relations between symbols and indirectly the fact of structural relations between occurrents. We have already seen that occurrents are united into complex occurrents through associative relations, and through the many kinds of dissociative relations including spatial distance and direction, temporal antecedence and consequence, spatial and temporal inclusion, etc. The problem of the operational origin of correlational symbols is that of the nature of the techniques involved in passing from the examination of such complexes to symbols whose meanings will then reveal the structural features of the complexes.

Before we plunge into this question, two remarks may be made with reference to the matter of intension as applied to indices and icons.

The essential feature of indices, we saw, lies in the fact that they have their content determined by the specific situation in which they are employed, and that they are, therefore, meaningless in isolation. They serve as directive factors, and not as characterizing symbols. There could be no significant way in which, through relations to other indices, the characters of the occurrents referred to could be specified. As a matter of fact most of the naming symbols in actual language have a certain intensional character; thus, for example, the symbol " this " is determined in its reference by its correlation with the symbol " that," and by its correlation with the symbol " these." Hence, it is actually possible to determine in advance something of the character of the occurrent referred to. But this is simply due to the fact that such symbols are not pure proper names. Even if there are not any pure proper names in language, there is at least this naming function which certain symbols possess and which certain other symbols lack. It is therefore correct to say that indices are strictly extensional, and have no significant relations to other symbols similar in character. Proper names cannot point indirectly; that which is being referred to must be given as the terminus of the act of pointing. Thus proper names do not unite to form symbolic systems, though they may and do as a matter of fact occur very significantly as elements in conceptual and propositional systems.

Icons have intension as well as extension. But there is clearly no specific problem here. If the function of the icon is that of representative portrayal, then the function of a complex icon must be that of portraying a complex occurrent. Hence, the intension of an icon is simply a further relational icon which aims to reveal the structural relation of the situation in question. Relational icons portray relational occurrents, and a complex icon consisting of relational and non-relational icons will portray a complex occurrent consisting of relational occurrents and occurrents thus related.

But the problem of the intension of characterizing symbols is much more complex. For here the attempt is made to represent the nature of the structure of occurents, but not to represent it in any pictorial manner. There may or there may not be a specific symbol which reveals this structural feature; often there is none, and the intensional aspect is indicated by mere juxtaposition. Furthermore, one must make a distinction between the symbol which describes the relation between the occurrents, and the symbolic relation between the symbols describing the respective occurrents. Thus we employ the symbol " association " to describe a certain kind of relation between occurrents, but the symbols which describe the two occurrents thus related are not themselves associated; association is a relation which holds only between occurrents, and the fact that this relation can be symbolized does not mean that the relation may also hold between symbols and so determine their structural relations. In the same way two occurrents may be adjacent in space, but we should not say that the symbol for the one occurrent is adjacent to the symbol for the other. And one occurrent may precede another in time or include another spatially, but we should not say that the symbol for the one precedes the symbol for the other, or includes it. A symbol for a relation is not necessarily a relation between symbols; it becomes such only when the correlational complex in which the two occurrents are united is a highly determinate one, and we can thus infer something of the character of the one occurrent from the character of the other and the relation. Hence, our specific problem is that of ascertaining a kind of relation between symbols which holds whenever the occurrents exemplifying the respective symbols are elements of a determinate correlational complex. For example, if a_1-b_1 represents an associative complex, what is the character of the relation between the symbol A and the symbol B by virtue of the fact that they are both exemplified in an associative complex? And if a_1 and b_1 are dissociated, is there any definite relation established between A and B by virtue of the fact that they participate, respectively, in dissociated situations? And does the repeated association or dissociation of occurrents determine any kind of symbolic relation between the respective symbols? This is the kind of problem we shall have to consider. As was pointed out at the close of the previous chapter, we shall not consider relations of temporal order or spatial separation, nor relations of spatial and temporal inclusion; we shall limit ourselves essentially to associative complexes.

Presumably the starting-point would be the analysis of a simple associative complex, such as a_1 -b_1. But with reference to such a complex we may note two things. First, it is an abbreviation of a more complex associative complex, which is a_1-s_1-b_1; where s_1 represents the situation. Secondly, since it is doubtful whether there ever is an intuitive awareness without thought, the more inclusive complex can be indicated thus:

$$a_1\text{-}s_1\text{-}b_1$$
$$\diagup\diagdown$$
$$\text{A} \quad \text{B}$$

where the fact of participation is indicated by the arrows. Then the fact of the participation of A and B in the situation is the correlative of the association of a and of b with the situation. If A and B participate, a and b are associated; and if a and b are associated, then A and B participate—or, at least if there is an intuitive awareness of a and b, then symbols such as A and B may be devised to complete the complex. The empirical theory of meaning insists upon the dependence of A and B upon a and b; the occurrents are fixed and determinative of the symbols which are flexible and determined.

Furthermore if we use participation as the basic fact, we have a convenient parallel form for the representation of non-participation. We must note that non-participation is not the equivalent of dissociation. Non-participation is the relation of a symbol to a situation with which *no* referent of the symbol is associated. Thus there is no method, by means of occurrents alone, for representing a generalized dissociation. The symbol $s_1 \leftrightarrow$ A may be used to represent the non-participation of A in the situation s_1; it implies that there is no a associated with the situation. There are three reasons for introducing the notion of non-participation. The first is the fact that the absence of an occurrent from a situation determines the character of this occurrent and of any occurrent present in the situation; hence, two occurrents united in this way constitute a correlational complex upon the basis of which inferences may be made to the respective elements. The second reason is the fact that an empirical theory of meaning will be obliged to show how such correlational symbols as incompatibility and contradiction have an occurrential foundation. But it is clearly impossible to show contradictory occurrents, for in the very nature of the case there can be no such occurrents; nor can one present an occurrential situation which exhibits incompatible occurrents, for they cannot be found in any given situation. Yet these logical relations have an empirical foundation. This foundation lies in the fact of participation and non-participation; by means of symbols we can represent a situation from which an occurrent is absent—something which cannot be represented without symbols. Hence, we can portray a situation illustrative of the non-participation of a symbol, and such a situation universalized affords the necessary origin for the symbolic relation of contradiction. The third reason is the necessity for giving meaning to negative symbols. Negative symbols are determined by a situation involving non-participation. If negative symbols referred to occurrents in the same way as positive symbols, they would presumably have their empirical origin in situations from which occurrents are absent. But these situations must be of a specific kind; i.e. they must be situations from which a certain kind of occurrent is absent. And the only way to represent the absence of a kind of occurrent from a situation is to employ the symbol for the designation of that occurrent and then to find the situation in which this symbol does not participate.

We are now ready to turn to a consideration of the most general of the relations determining the structural content of symbols. This is the occurrential relation of otherness, which is determinative of the symbolic relation of diversity. This is not an important correlational symbol, for it is not highly determinative and there is therefore no operation which is based upon it. But it must be examined in order to see how the diversity of occurrents is reflected in the diversity of symbols.

Diversity

We have seen that the plurality of the realm of occurrents resides in the fact of the relation of otherness between any two occurrents. This relation holds when any two occurrents are associated with the same space and time but qualitatively different, or associated with different spaces or times. Otherness is spatio-temporal or qualitative, though it may be both. Now the adequacy of the symbolic system demands that the plurality of occurrents should have a counterpart in the plurality of symbols. But the parallelism cannot be complete. For it is of the very essence of descriptive symbols that their reference is through generalization, and generalization involves loss of particularity. *Thus one cannot refer uniquely through a general symbol.* An alternative statement of the same fact is the assertion that the " here " and the " there " in space and time cannot be descriptively referred to. Consequently the demand that the plurality of the realm of occurrents be reflected in the realm of symbols cannot be met completely. The only possible plurality in the realm of symbols will be a qualitative plurality. This means that differences in symbol will reflect differences in *kinds* of occurrents but not in individual occurrents. Occurrential otherness will not imply symbolic otherness, though symbolic otherness will imply occurrential otherness. But a *qualitative* occurrential otherness will have its counterpart in a diversity in the symbolic system; different kinds of occurrent will be represented by symbols with different unique contents. It seems necessary, therefore, to suppose a relation of general qualitative distinctness between symbols, which is derived from and patterned after the occurrential relation of qualitative otherness.

That such a relation does exist in any system of descriptive symbols is obvious. The symbols are diverse from one another, and the diversity is indicative of diversity in referents. There can be no strict synonyms in a symbolic system; for if the symbols have different linguistic representations, and the linguistic representations are uniquely associated with the meanings, then there must be qualitative diversity in the referents. Otherwise our symbolic system is redundant.

Compatibility and Non-implication

Much more significant in the determination of the structural relations of symbols are those which are based on the existence of associative and dissociative complexes of occurrents. But the starting-point here, as was pointed out a moment ago, is not the associative or the dissociative complex but the participation and the non-participation of symbols in situations. The basic principle of symbolic structure is that expressed in the fact that situations may exhibit various combinations of participation and non-participation of symbols. Since we are, for the present, concerned primarily with the simplest of these relations, let us limit our considerations to the possible references of two symbols to a situation. Given two symbols both may participate, one may participate and the other may not, or neither may participate. Each of these situations involving participation or non-participation determines a relation of the symbols to one another, though the relation determined by joint non-participation is of no essential significance in the building of systems of symbols, since it is inexpressible without the employment of negative symbols, and these, we shall see later, have their meanings determined by relations to other symbols which participate directly. But all of the other three determine relations of significance. They are as follows : Joint participation determines the symbolic relation of *compatibility* or *conjunction* ; participation of one symbol with non-participation of the other determines that the symbol descriptive of the one is *non-implicative* with reference to the other ; participation of the other symbol with non-participation of the one determines that the other symbol is *non-implicative* with reference to the one. Thus we can say that if any two occurrents are associated, the symbols which characterize them respectively take on a relation to one another which we express by saying that they are compatible or conjoined with one another ; if any occurrent is found in a situation in which there is no occurrent of a specified kind, then the symbols which characterize them respectively take on a relation to one another which we express by saying that the first does not imply the second.

This fact of the structural inter-relation of symbols is of great importance, though the instances of such structural relations which we have considered thus far are among the least important. Two points, however, should be called to attention. (1) It is through such relations as this that symbols take on organization with reference to one another and ultimately unite into symbolic systems. A symbolic system is a group of symbols each of which has definite structural relations to the others. But since these symbolic relations have been derived from and are therefore reflections of relations of occurrents, we see how it is possible to construct a symbolic system descriptive of an inclusive occurrential realm. The ultimate goal of knowledge is the construction of a symbolic system whose elements represent through characterization the individual occurrents, and whose structural relations represent through correlation the associational and dissociational relations which constitute the organization of occurrents. (2) The intensional character of a symbol which is related structurally to another symbol determines an indirect reference of the symbol in question to the realm of occurrents. Every symbol, then, by virtue of structural relations possesses a double reference to occurrents. There is the direct relation which the symbol bears to the occurrents which are its examples; this relation, or manner of referring, holds even in the case of symbols which are not known to be exemplified at all. Such symbols cannot have been derived empirically but must have been derived from other symbols; nevertheless, they exhibit the form of referring to occurrents. But there is also the indirect relation which each symbol bears to occurrents by virtue of its structural relations to other symbols which themselves have direct reference to occurrents. Thus if two symbols A and B are compatible, A refers to a_1 directly and B refers to b_1 directly, but A also refers indirectly to b_1 and B refers indirectly to a_1. This fact is of great importance, *for the very possibility of using meaningful symbols prior to the knowledge of situations in which they participate is dependent on it*. Correlational symbols determine operational routes by which the symbols for obscure occurrents are given precise content. If the correlational symbol represents the structural relation between two occurrents, one of which is given clearly and the other of which is given obscurely, then by means of this symbol and the content of the clearly given occurrent we can define the content of the obscurely given occurrent. If the relation is highly determinate, we may ascertain in advance of any clear awareness of that occurrent many of its features; but even if the relation is not so highly determinate as to content, it will usually be determinate as to location, thus we shall know at least where to look for the obscure occurrent. Hence, this fact of the indirect reference of symbols makes possible not only the legitimate use of hypotheses and other suppositional symbols but also acts of verification.

We may illustrate the establishment of relations of compatibility and non-implication between symbols in terms of the two main types of characterizing symbols,—viz., concepts and propositions. We have seen that these are both types of universalized characterizing symbols. However, the former refer to occurrents as unanalyzed, while the latter refer to them as analyzed; and the former usually imply no existential claim as part of their manner of referring, while the latter make assertions.

It will be advisable to call attention in advance to the peculiar inter-relation of these two types of symbol. On the one hand we may consider concepts as our symbolic elements; then the problem of the determination of symbolic structure is that of determining the structural relations of concepts to one another. Hence, the intension of a concept will be the totality of its structural relations to other concepts. This is broader than the ordinary notion; usually it includes only the implicates of a given concept. But what is important here is that the correlational symbol connecting the concepts is a representation of the relation between the respective occurrents. And we saw in chapter IV that in all cases the associative union of occurrents determines a new occurrent. Thus the separate occurrents unite to form a more inclusive occurrent, and we should also expect that the separate concepts should unite to form a more inclusive concept. This is precisely what takes place; the new concept is the conjunctive union of the separate concepts. Sometimes a new symbol is invented to designate it; but in general it is represented by the separate symbols connected by the word " and " or its equivalent. Usually the new symbol, in case one is constructed, is indefinite as to content, since it aims to portray not merely the two occurrents given but all occurrents to be found in the associative complex; these are not given for awareness, hence the content of the concept must be made sufficiently flexible to include them when they are discovered. (When we are concerned with repeated occurrents, this same flexibility is found; but here the content of the newly devised concept is progressively increased by the discovery of new associated occurrents. This seems to me to be the explanation of the difference between analytic and synthetic propositions; all propositions describing the character of associational complexes are synthetic in their first formulation because the content of the subject concept has not yet been determined, but are analytic ever after because the subject concept then takes on as part of its implicates the predicate just attributed to it).[1] Hence in any associative union of two occurrents, say a_1-b_1, there is to be found the empirical basis for the relation of compatibility between the concepts A and B, and also for the formulation of a new concept, C, descriptive of the associative union of the two concepts.

We may refer again to a point mentioned in chapter III. The fact that c designates a complex of which a and b are elements determines a relationship of relative concreteness and abstractness between c, and a and b respectively. Thus we tend to look upon c as a "thing" which possesses a and b as attributes. This is clearly seen in the fact that in the mere associative complex a_1-b_1, which has not been conceptually characterized, we find that either element may be considered the "thing" of which the other becomes the attribute; e.g., if it is the association of a red and a triangle, we may say either that the red is triangular or that the triangle is red. But when the associative complex is clearly recognized and a term has been devised to symbolize it, we see that the accurate description of the totality demands that it be only the red-triangle which is red and triangular. The relativity of the concreteness and abstractness of the whole and the parts can be seen in the fact that a more inclusive associative union may be later discovered; it will then be desirable to express the complex red-triangle as an attribute of this whole, in which case it will become abstract relative to the whole. For example, pitch and volume may both be attributes of a tone, which is the associative union of them; but the tone may be an attribute of a violin which produces it; thus the tone is concrete relative to its volume, which is abstract, but the tone is abstract relative to the violin, which is concrete.

Another point which may be called to attention here is that since the associative union of two occurrents determines a new occurrent which is the associative union of the two, and which is itself associated with each of the two, there arises the necessity for devising a number of alternative but equivalent symbolic expressions. If red and triangle are associated to form a red-triangle one expresses the fact symbolically by saying that the red-triangle is red and that the red-triangle is triangular.

[1] See chapter IV, p. 75.

This, alone, is of no special significance. But it has an important consequence. By means of it we may see in what sense diversity in the contents of two concepts necessitates diversity in their extensional references. There is clearly a sense in which red and triangle both refer extensionally to the same occurrent,— viz., the complex occurrent which is the associative union of red and triangle. But there is another sense in which there is diversity of extension. The occurrent which is red (i.e. the red occurrent) is not identical with the occurrent which is a triangle (i.e. the triangular occurrent). Thus diversity in content demands diversity in extension. But suppose one means to refer in the extension to the complex occurrent of which the occurrent properly designated in the meaning of the concept is an element. It is commonly said, for example, that equilateral triangle and equiangular triangle are identical in extension. This is true if one refers to the complex occurrents which are the triangles, but it is not true if one refers to the occurrent which is the equality of the angles as over against the occurrent which is the equality of the sides. If two concepts were strictly identical in extension, they could not possibly differ in meaning; for meaning is determined simply by a given operation upon a given occurrent, and similar generalizations upon a given occurrent could not possibly produce divergent meanings. This same point is applicable to occurrents which would ordinarily be called species and genus with reference to one another. If red is associated with color (from which it follows that color is associated with red), one may say that the red is colored, or that the color is red, or that the red-color is red or that the red-color is colored. Here red and color refer to the same associative complex,—viz., the red-color. But they do not have identity of extension, even in a given occurrent. Thus when concepts are related as species to genus, the concepts which are their respective exemplifications are diverse, though they may be (and, in fact, are) associated in at least one situation. Certainly color does not mean the same as red so that the two concepts cannot be identical in content. And when we say that they are identical in extension we mean simply that one cannot point to a red occurrent without also at the same time pointing to a colored occurrent. But this does not argue for their extensional identity ; it argues, rather, for the vagueness of pointing when the referents are abstract occurrents instead of concrete occurrents. Concepts are genuine principles of individuation.

So much for the conceptual interpretation. The propositional interpretation will differ from this in two ways : (1) It will indicate more clearly the structure of the complex occurrent which is being referred to. (2) It will contain a definite assertive reference to such occurrent. With regard to the former, the propositional interpretation of the joint participation in a given situation will be precisely the assertion of the compatibility of the concepts thus participating. For the natural manner of representing the intension of a concept is the assertion of a proposition connecting this concept with another. Thus when we assert that A is compatible with B, we are asserting the participation of a propositional symbol in a_1-b_1, and the propositional symbol clearly indicates this fact. In the propositional representation we retain the structure, whereas in the conceptual representation we tend to lose sight of the structure. With regard to the assertive aspect, when we employ the proposition " A is compatible with B," we consider the assertive claim as part of the meaning of the proposition. Thus such a proposition claims the existence of the occurrent, a_1-b_1, while the concept C as the conjunctive union of A and B makes no such claim ; i.e., the concept C could be meaningful even though it were devised to designate the conjunctive totality of two concepts not known to participate jointly in any situation. But in the propositional symbol the existential claim is more intimately tied up with the meaning.

It is helpful also to consider the interrelation between concepts and propositions with reference to the distinction between unique and structural contents of symbols. The determination of the unique content of a symbol is made by referring to the situation in which the symbol participates ; there will always be found in such a situation the specific occurrent which is the example of the symbol, and from this the unique content of the symbol may be determined. The structural content of the symbol is then determined by noting with what occurrents the occurrent in question is associated and dissociated. In this way we determine structural content with reference to occurrents which are structurally related to the original. But now consider the respective processes for the determination of the unique and structural contents of concepts and propositions. To determine the unique content of A we refer to a_1, and to determine the structural content of A we refer to b_1, which is an occurrent associated with a_1. Then we express the result of our investigation by saying that A is compatible with B. But this is precisely a propositional symbol descriptive of the situation. Thus the process for determining the unique content of the proposition " A is compatible with B " is the same as that of determining the structural content of one of its elements. *Hence, the unique content of a proposition always reveals the structural contents of its elements.* But the proposition itself has a structural content; it is the relation of the proposition to all other propositions. This is determined by noting what other propositional symbols participate and do not participate in the occurrential situation a_1-b_1. For example, if the occurrential situation could be shown to be $(a_1$-$b_1)$-$(c_1$-$d_1)$, then the proposition " C is compatible with D " would also participate in the situation, and one could then assert that the proposition " A is compatible with B " is compatible with the proposition " C is compatible with D." This means in general that if we look at a proposition we can see it both as a symbol for the structural organization of occurrents and as an element of a larger structural organization. And if we look at a system of propositions we can see it as a symbol for the structure of structure. This is important, for it enables us to see the empirical foundation of *inference*. However, the topic is too broad to be discussed here. I have tried only to indicate the strict parallelism between the realm of occurrents and the realm of symbols. To simple occurrents there correspond simple concepts; to complex occurrents (i.e. associative and dissociative unions of simple occurrents) there correspond either complex concepts or simple propositions made up of the interrelation of the concepts for the simple occurrents; to complex complex occurrents (i.e. associative and dissociative unions of complex occurrents) there correspond either highly complex concepts or compound propositions made up of the interrelation of simple propositions for the complex occurrents. Except for very simple occurrents there is always a double interpretation : one conceptual, which does not clearly reveal the structure, and the other propositional, which does.

We have examined the relation of compatibility; let us now consider the relation of non-implication. This, we saw, is the relation between one symbol and another when the former participates in a situation in which the latter does not. What determines this relation between symbols is not merely the fact of dissociation, but the fact of a partially universalized dissociation. If two occurrents are dissociated, they are associated with different situations; but they may be either qualitatively alike or qualitatively different. Hence, from the mere fact of the dissociation of two occurrents one could not determine any significant relation between their respective symbols; for if the occurrents are qualitatively alike, then the relation between the symbols would be a relation of a symbol to itself. Consequently the statement, " This A is not associated with that B," tells us nothing of importance about either A or B.

But the statement of the respective participation and non-participation of two symbols in a situation does assert a structural feature of the world, although its negative character makes it less highly determinate than the corresponding positive correlation. If I state, " This A is not associated with *any* B," I learn something of importance about both A and B,—viz., that B cannot be dependent upon A. Hence, I establish the relation of non-implication from A to B. Such a relation is part of the intensional content of each of the symbols, and the fact that it is neither positive nor highly determinate should not obliterate this fact. To know what a symbol does not imply and to know what does not imply a given symbol are both bits of information which may in certain cases prove to be important.

We have thus far discussed three kinds of correlational symbols,—viz., diversity, compatibility and non-implication. None of these is highly determinate. If we know of a certain concept, say A, that it is diverse from B, compatible with C, and non-implicative of D, we know very little about it. But we know something. For example, if C also implied E, we could then conclude that A is compatible with E; also if D were implied by F, then A would not imply F. Since the relations of compatibility and non-implication are expressed by the ordinary "I" and "O" propositions of the Aristotelian logic, the conclusions here given can easily be justified. But they are obtainable only when combined with relations expressible in the "A" and "E" propositions; hence, the relations of compatibility and non-implication are not highly determinative (i.e. they do not permit an extensive determination of content). Thus if A were a symbol for an obscure occurrent, and we were endeavoring to give it content through an operational route based only upon such relations, we should meet with little success. But this should not obliterate the fact that these are legitimate correlational symbols, revealing the intensional character of symbolic elements and thus giving structure to the total realm of symbols.

Co-implication, Implication, Incompatibility

Much more significant among correlational symbols are those which are based upon *repeated associations* and *repeated dissociations*, or, in the alternative phraseology, *repeated joint participations* and *repeated participations and non-participations*. The former determine the symbolic relation of *implication*, and the latter of *incompatibility*. Since these constitute a sub-type of what are generally recognized as *scientific laws*, let us stop a moment and examine this notion. In order to indicate that this type of structure is actually that which is revealed in scientific laws, let me quote several authorities as to the meaning of the notion in science. " Laws are propositions asserting relations [of uniform association] which can be established by experiments or observation."[1] " A natural law. . . . is essentially a general correlation between classes."[2] " The ' Laws of Nature ' are descriptive formulae in ' conceptual shorthand ' of the routine of perceptions."[3] A law states " an invariable relation of succes-

[1] Campbell, *Physics, the Elements*, p. 38.
[2] Ritchie, *Scientific Method*, p. 62.
[3] J. A. Thomson, *Introduction to Science* (New York, 1911), p. 56.

sion or of coexistence."[1] " The expression, Laws of Nature, *means* nothing but the uniformities which exist among natural phenomena . . . when reduced to their simplest expression."[2] These are but a limited selection from a large number of assertions of a similar character.[3] Thus in discussing repeated structures we are attempting to ascertain the empirical foundation for laws.[4]

The exact statements of the occurrential structures at the basis of these symbolic relations are difficult to make. Roughly one may state them as follows : (*a*) If whenever a symbol, A, participates in a situation another symbol, B, also participates, then A implies B. (*b*) If whenever B participates A also participates, then B implies A. (*c*) If both conditions are satisfied, then A and B are co-implicates ; i.e. A implies B, and B implies A. (*d*) If, whenever A participates, B does, but B sometimes participates when A does not, then A implies B, but B does not imply A. (*e*) So also conversely. (*f*) If, whenever A participates, B does not, and, whenever B participates, A does not, then A is incompatible with B, and B is incompatible with A. The conceptual interpretations of these structures as given by the Aristotelian logic would then be as follows : (*a*) All A is B. (*b*) All B is A. (*c*) All A is B and all B is A. (*d*) All A is B and some B is not A. (*e*) All B is A and some A is not B. (*f*) No A is B and no B is A.

[1] F. Enriques, *Problems of Science* (Chicago, 1914), p. 68.
[2] J. S. Mill, *System of Logic* (London, 1868), Vol. 1, p. 354.
[3] Cf. also Whitehead, *Adventures of Ideas* (New York, 1933), chap VII.
[4] The important question as to whether general laws can be validly derived from particular associations and dissociations is not to be examined here. In other words I shall not be concerned with the *justification* of generalized correlation. There are some serious doubts in my own mind that generalization has any logical justification. If it is an operational route from occurrents to symbols it can have no *formal* justification for no such route can be formalized. The attempts of Keynes (*Treatise on Probability* (London, 1929)), and Nicod (*Foundations of Geometry and Induction* (London, 1930)) to found it on probability are certainly more hopeful than Mill's attempt to make it universally valid. But we may be obliged to conclude that we can merely describe how we proceed in such cases and hence that we must admit only a " prescriptive rule." For the concise statement of this view see H. Feigl, " The Logical Character of the Principle of Induction," *Philosophy of Science*, Vol. I, pp. 20–29. In Chapter IX I shall include generalization among the hypothetical rather than the constructional routes for giving contents to symbols.

The more precise statement of the occurrential structures determining the correlational symbols of implication and incompatibility can be made if we emphasize the notion of class. A class was defined[1] as a homogeneous complex occurrent,— i.e., the dissociative union of several occurrents having a qualitative similarity. Then participation in such an occurrent must be defined in a specific way, depending upon the extent of the homogeneity. A class of complex occurrents may be completely homogeneous with respect to one element of each of the complexes but only partially homogeneous with respect to the other. For example, a group of men may be completely homogeneous with respect to the attribute of humanity, but only partially homogeneous with reference to whiteness of skin. Then, given an extended homogeneous occurrent made up of complex occurrents and of such a character that it is completely homogeneous with respect to the participation of B but only partially homogeneous with respect to the participation of A, A implies B but B does not imply A. Given an extended homogeneous occurrent in which there is complete joint participation of symbols A and B, then A and B are mutually implicative. Given a homogeneous occurrent in which a symbol A completely participates and a symbol B does not participate at all, and another homogeneous occurrent in which a symbol B completely participates and a symbol A does not participate at all, then A and B are incompatible. But since these homogeneous occurrents are too extensive to enter readily into awareness, we must infer their characters from samples taken from them. Hence, we may say that if sample a's are of the kind A $\leftarrow s_1 \rightarrow$ B, and sample b's are of the kinds A $\leftarrow s_1 \rightarrow$ B and A $\leftarrow+ s_2 \rightarrow$ B, then A implies B. But if sample a's and sample b's are both of the kind A $\leftarrow s_1 \rightarrow$ B then A and B are mutually implicative. And if sample a's are of the kind A $\leftarrow s_1 +\rightarrow$ B and sample b's are of the kind A $\leftarrow+ s_2 \rightarrow$ B, then A and B are mutually incompatible.

The operation of symbolic derivation in this case is one of repeated correlation, and is thus different in character from that involved in the derivation of simple correlational symbols such

[1] Chapter IV, p. 81.

as compatibility and non-implication. There the generality of the symbol referred merely to the fact that it was applicable to a range of occurrents possessing a certain qualitative similarity to the occurrent from which it was derived; it expressed the fact that, if there were such occurrents, they were also capable of being represented by the symbol in question. But here there is not merely a reference to other situations similar in character. A universal or general proposition does not mean simply that whenever there is a complex aRb it may be considered as the referent of the proposition; it asserts universality in a different and more far-reaching sense. It affirms that whenever an occurrent bearing a certain qualitative similarity to one element of an associative or dissociative complex is found, there will also be found in the proper associative or dissociative relation an occurrent bearing a certain qualitative similarity to the other element of the complex. More simply, but less precisely, it asserts that complexes tend to recur, that the associative and dissociative relations of an occurrent to other occurrents are parts of the essential contents of the occurrents involved, and, hence, that the recurrence of an occurrent (i.e. the happening at a different space or time of a highly similar occurrent) necessitates a recurrence of its structural relations to other occurrents. An occurrent is not an isolated entity but a participant in a large number of relational complexes. So far as we can tell, occurrents of certain kinds always are to be found in such complexes. A universal proposition asserts not merely that there are more of such complexes than we have observed, but also that the complexes which we have observed give us the *type* of all such complexes.

The result is that universal correlational symbols are highly determinative of the structure of occurrents. If we know of a certain obscure occurrent that its symbol can be defined as the co-implicate or the implicate or the implicans of the symbol for a clearly given occurrent, we know with a relatively high degree of determination both where the obscure occurrent is to be found and what its content is. Thus we have an invaluable instrument for defining the symbol referring to the obscure occurrent, in advance of any clear awareness of that occurrent.

In this way we can talk about it before we know whether it exists, or, if it exists, what character it must possess. But we are not thereby talking about it in a random manner; we know that if our symbolic analysis of the realm of clear occurrents has been correctly performed, then the occurrent in question *must* exist where we predict it to exist and it *must* have precisely such content as we infer it to have. Of course it may have more content than we know it to possess in this advance awareness, and we hope that it will, for only in such a way can scientific knowledge increase. Also the occurrent in question may be found not to exist, or an occurrent having a distinctly different content may be found to exist in its stead; but we then conclude either that our symbolic analysis of the realm of the clearly given has not been correctly made or that our operational route from the given has been erroneous. Hence, the ultimate criterion of symbolic applicability is always the realm of occurrents. But this does not prevent us from ascertaining in advance of awareness what the realm of occurrents when explored in a given locus is likely to reveal.

The two correlational symbols, co-implication and implication, determine, by virtue of the asymmetric character of the latter, three operational routes. These vary in the degree of determinateness. Passage from a symbol to its co-implicate is the most highly determinate of operations, for by means of it one can transfer all of the structural relations of the one symbol to the other, and one can also transfer its existential claim. Thus if A participates in a situation s_1 and B is a co-implicate of A, then we know that B participates in the situation and that all of the structural associations and dissociations of a_1 will also be structural associations and dissociations of b_1. But if A participates in a situation s_1 and B is an implicate (but not an implicans) of A then although we know that B participates in the situation we know less about the structural content of b_1; for example, the implicans of A may be transferred to B as its implicans but the implicates of A may not be transferred to B as its implicates. And if A participates in a situation s_1 and B is an implicans (but not an implicate) of A, then neither do we know that there is an occurrent b_1 in the situation, nor are we able to ascertain much of the content of a b occurrent. For example, the implicates of A may be transferred to B as its implicates but the implicans of A may not be transferred to B as its implicans. Relatively speaking, there is a higher determination of content in this case than in the preceeding, but the loss in existential claim makes it a less useful operational route.

The reasons for these differences in the determination of correlational symbols can be seen by referring to the occurrential structures of which they are a reflection. The distinction between concrete (complex) occurrents and abstract (simple) occurrents, to which we referred in chapter III, is determined by the fact of repeated associations. Simple association (i.e., the occurrence of an isolated associative complex) cannot determine such a distinction, for it is a symmetrical relation and the respective terms of the relation are interchangeable. But if we have a repeated association of such character that an exemplification of A is always found in association with an exemplification of B, but an exemplification of B is sometimes found in dissociation from any A, then the relation of A to B cannot be the same as the relation of B to A. Let us express this by saying that the symbol A is concrete with reference to the symbol B, and that in any situation, a_1-b_1, a_1 is concrete with reference to b_1. Now this is the expression of a structural relation between occurrents, and it can be shown that it is a more satisfactory definition of these notions than that introduced in chapter III, where we defined concrete occurrents as those which include occurrents (in the associative sense), and abstract occurrents as those which are included in occurrents. The inadequacy of this definition lies in its inapplicability; because of the inherent obscurity of complex occurrents and the tendency to confuse them with one or more of their associated elements, one cannot tell by the inspection of a situation whether the relation in question is one of inclusion (i.e. concrete to abstract) or association. One distinguishes an inclusive occurrent only when a *repeated* associative complex is noted, and the devising of a symbol for its portrayal necessarily involves a recognition of its inherent generality, i.e., its repeatability. Thus in order to establish the fact that an inclusive occurrent, say an F, is always found as including the sub-occurrents, say an A and a B, we must establish the existence of a repeated associative complex, $f_1\text{-}a_1\text{-}b_1, f_2\text{-}a_2\text{-}b_2, f_3\text{-}a_3\text{-}b_3$. . . . In other words, in order to recognize that f_1 includes a_1 and b_1 one must note that an F always occurs in association with an A and a B. It can thus be shown that the associative tie is *necessary* for the determination of relative concreteness and abstractness; for whenever a whole occurs the parts must occur, but one part may occur in the absence of the whole. It can also be shown that the associative tie is *sufficient* for the determination of relative concreteness and abstractness, for that which distinguishes an abstraction from its corresponding concretions is precisely the fact that the abstraction is more general, i.e., occurs in a wider range of situations than the concretion and thus may occur in the absence of the concretion, while the concretion occurs in a more limited range of situations and thus is never found in the absence of the abstraction. The relation of concrete to abstract is the essence of the relation of species to genus, except that the latter usually implies an essential similarity between the symbols thus related.

The correlational symbol of incompatibility is relatively indeterminate, and involves a complete loss of existential claim. Its indeterminateness resides in its negative character. If we have a clearly given occurrent to which an obscurely given occurrent, defined through its symbol, is related by incompatibility, then only the affirmative structural relations of the former may be transferred, and they must be attributed to the symbol for the obscurely given occurrent as negative relations. None of the negative structural relations of the original may be transferred. Thus if this route is employed for the purpose of defining the symbol for the obscure occurrent, it is quite inadequate; we know nothing about what the occurrent is and very little about what the occurrent is not. If we add to this the fact of the loss of existential claim, it becomes clear that an operational route along the structural relation of incompatibility is of no essential aid in furthering the development of knowledge; it is often important to know what is not present in a situation, but it is usually more important to know what is present, and it is always important to know with the highest possible determination of content what is suspected to be either present or absent.

Contradiction

There is one final relation between symbols which must be here examined; this is the relation of contradiction. Here the empirical derivation cannot be of the same direct sort as the correlational symbols thus far considered,—i.e., there can be no occurrential relation of contradiction from which it can be derived by generalization. The situation is not quite the same as in the case of incompatibility. Apparently incompatibility can be derived from empirical situations exemplifying it. Of course, this is really not the case; there are no instances of empirical incompatibility. The foundation of the proposition "A is incompatible with B" was shown to be a class of occurrents some of which were characterizable by A but not by B and the rest of which were characterizable by B but not by A. However, we found that we could examine the former cases and note the absence of B, because we knew *from other situations in which B participated* what B was; and we could examine the former cases and note the absence of A, because we knew *from other situations in which A participated* what A was. Hence, we could determine the incompatibility of A and B because we had been given A and B and had discovered that they never occurred in conjunction.

The case is not quite the same for contradiction. A is given but there is no way in which non-A can be given; hence there is no way of determining whether it is present or absent in a given situation. The empirical foundation of contradiction cannot be the participation of A and the non-participation of non-A in a situation, for the content of non-A is determined precisely by the fact that it does not participate in a situation in which A does participate. However, it seems to be the case that in every situaton in which there is a participation, there is also a non-participation; and in every situation in which there is a non-participation, there is also a participation. That is, every situation which seems to be only positive in character proves upon examination to be also negative in character, and every situation which seems to be the mere absence of a symbol proves upon examination to exhibit a participation. Contradiction has its empirical foundation in this fact. It arises out of the following:

(*a*) Every situation which exemplifies P fails to exemplify Q. (*b*) Every situation which fails to exemplify P exemplifies Q. This makes the symbols P and Q contradictories of one another. But since non-exemplification is always a derivative fact (i.e., one can discern non-exemplification only if one knows in advance the character of the symbol which is not exemplified) statement (*a*) must be employed before statement (*b*) in the definition of Q, and statement (*b*) before statement (*a*) in the definition of P. In other words, I am given P first as exemplified; from this I define Q as the symbol which is not exemplified in this situation. Then, knowing what P is, I discover a situation in which P is not exemplified. I then further define Q as that which is exemplified in this situation. This makes P, participation, and non-participation the indefinables, and defines Q in terms of them. In order to show that Q is operationally derived from a given concept, it is advisable to designate it by a specific symbol; let this be non-P. Then non-P may be defined as that which fails of exemplification in all situations in which P is exemplified and is exemplified in all situations in which P fails of exemplification. Here the order of the defining properties stated in the definition gives the order of operational procedure.

It is clear that the operational route defined through this complicated structural relation is not highly determinate; negatives are essentially void of specific content, for they are defined through a kind of generalized otherness. Hence it is dangerous to think with such symbols, for they are ambiguous until their operational routes have been determined. But even when their contents are thus determined, they remain partly ambiguous, for one can never be sure as to the scope of the universe of discourse,—e.g., whether the negative operation employed in the determination of " non-white " is based upon the participation of a specific white, a more abstract white, color in general, visual properties as a whole, or some even more highly abstract symbol. Thus one cannot tell whether " non-white " is to be defined as all whites other than the given one, all colors other than white, all non-colors, all non-visual properties, or the denial of some more highly abstract symbol. For this reason, and because of the loss in existential claim when these symbols are employed, they are best considered simply as symbolic constructions employed to further the reasoning processes but not supposed to enter either into the original data or into the final conclusion. They are tools or devices to be abandoned in the final structure, and can be meaningfully employed as tools only if their contents are defined through operational routes from symbols which are themselves well defined and refer directly to clearly given occurrents.

Two features of the relation of contradiction may be briefly referred to. The first is the fact that it can hold only between characterising and correlational symbols. Since it is found only through the operation of generalization neither names nor icons may be contradictories of one another. If one designates an occurrent as " not-this " he is clearly referring to a class of occurrents which are the others of this occurrent, and such a reference to a class must take place through characterizing symbols. So also if an icon fails to portray an occurrent one ought not to say that the icon is the negative of another icon which does portray the occurrent; the negative of an occurrent is a class of occurrents differing from the given occurrent in all possible ways; hence to speak of the negative of an icon is to suppose it a characterizing symbol possessing generality of reference. An icon is a particular, just as is the occurrent referred to; hence the relation between two icons is the same as that of any two differing occurrents. The second point to be called to attention is the relation of the notion of contradiction to the fact of positive and negative reference, which we found to be a feature of all general characterizing symbols. Every such symbol refers to the total realm of occurrents but it refers to a part of that total realm (viz. the occurrents which are its direct referents) in a positive way, and to the other part (viz. the occurrents which are its indirect referents) in a negative way. Thus every such symbol possesses a double meaning or a double way of referring. Contradiction is the relation which holds between any two symbols when each means the direct referents of the other in an indirect way and the indirect referents of the other in a direct way. The operation of contradiction is thus a means for transferring the general manner of meaning into the specific content of a new meaning; i.e. if A means certain occurrents negatively, then a newly created symbol, non-A, will mean these same occurrents positively, and conversely. Thus A and non-A are distinct symbols, having a generalized characterizing reference to occurrents and so related that they cannot both refer positively to the same occurrent yet must disjunctively refer to the total realm of occurrents.

The results of this long chapter may now be summarized. We have attempted to show the empirical foundation and the operational routes for the derivation of correlational symbols. Correlational symbols portray the structure of the realm of occurrents, but they reveal this structure in an indirect way by referring to a relation between symbols which they reveal in a direct way. When occurrents are structurally related in certain ways which permit us to say that they constitute correlational complexes, these structural relations are capable of being symbolized through correlational symbols. The most important of these are otherness, compatibility, non-implication, co-implication, implication, incompatibility, and contradiction. All of these designate ways in which occurrents are related to one another, though in certain cases we determine the meanings of the symbols not by an examination of the complexes but by noting instances of joint and disjoint participation. Since these designate kinds of structural relations between occurrents, they become invaluable instruments by which the contents and existential status of obscurely given occurrents may be determined. Some of the relations permit a high degree of determination; others permit only a relatively low degree. They thus define operational routes by which in advance of a clear-cut awareness of specific areas of the realm of occurrents we may determine what is to be found. They make possible *meaningful* anticipations of experience.

Before passing to the next chapter it will be advisable to suggest the direction which a more complete discussion of the present topic would take. We have limited ourselves to the structure of associative complexes, and within these structures we have noted only two correlational symbols exhibiting a relatively high degree of determination,—viz., co-implication, and implication (in its two directional variations); we thus have an operational technique for defining from a clearly given occurrent its co-associates, its abstractions, and its concretions. Now, however significant these are, they are only a few of the important structural relations of occurrents.[1] We have already seen that relations of temporal succession, spatial and temporal inclusion, spatial proximity, etc., are also significant. Thus a complete discussion of this topic should include the operational routes for determining the meaning of symbols for the past and the future from the present, the meaning of symbols for wholes and parts from middle-sized objects, and the meaning of symbols for adjacent spatial objects from objects which are here. This would involve a consideration of temporal (causal) laws, rules for analysis and synthesis, and principles of spatial structure. Since these considerations would extend the present volume by many pages, I must remain content with a mere mention of the type of problem involved. A complete operational theory of the development of scientific concepts would demand solutions to these problems.

---

[1] Thus in opposition to R. D. Carmichael (*Logic of Discovery* (Chicago 1930), pp. 21–27) I should insist that there are different logics of discovery for different subject matters, though there may be a common structural feature prevading all subject matters. This would presumably be the structure of the abstract-concrete relations.

CHAPTER IX
MEANING: CONSTRUCTS AND HYPOTHESES

The topic to be discussed in this chapter is without question one of the most important features of the structure of science. In order to make this fact clear let us summarize the results thus far attained and endeavor to determine what the logical structure of science is without mention of this highly significant aspect, i.e. what we should be obliged to conclude as to the character of science if we should terminate our discussion at this point.

Symbols arise through operational consideration of the realm of occurrents, and as a result take on meaning properties which are expressive of the occurrents characterized and of the operational routes employed in the derivation of the symbols. There are symbols for occurrents themselves, and symbols for the structures of occurrents. The former arise through operational activity upon isolated occurrents and endeavor to portray their unique contents; the latter arise through operational acts performed upon associative and dissociative complexes and aim to reveal the organization of nature. Science as a whole is an activity of confronting oneself with a world of occurrents, and directing upon this world certain operative acts whose purpose is the formation of symbolic representatives having a perfect isomorphism with the given world.

The most superficial examination of any science convinces one of the insufficiency of this analysis. Clearly one of the most important features of science is that *it talks about things which do not obviously exist in order to convey information about things which do obviously exist*. No discussion of science is complete without a recognition of the fact that the scientist not only often talks *in advance* about occurrents, but also sometimes talks about them through reference to entities which may even be incapable of existing. Hence mention must be made of hypothetical and theoretical entities, fictional elements, constructions and idealizations.[1] Whether these entities constitute a part of the genuine

[1] For a list of terms characterizing such entities see Vaihinger, *Philosophy of 'As If'* (London, 1925), pp. 96–7.

subject-matter of science, or whether they are merely "picturesque and interesting" ways of talking about nature,[1] are questions which we shall have to consider. But in either case they are at least found in science, and we shall be obliged to examine their claims.

The omnipresence of such entities in science makes any preliminary listing of them in the interests of illustration impossible. Hypotheses, for example, cannot be listed because such entities have a more or less temporary status and pass very soon into the realm of the clearly given occurrents; what is an hypothesis at one time may be fact at a later time. It is perhaps not far from the truth to say that since the discovery of the Brownian movement molecules have ceased to be hypothetical. But one can say, subject to this recognized inadequacy, that atoms, electrons, ether, vital forces, souls, etc., are hypotheses. Differing from these in ways which we shall examine in this chapter are constructs. In this group we find such entities as perfect points, lines, and figures, negatives, irrationals, imaginaries, infinitesimals, perfect levers and gases, absolute motion and rest, perfect rigidity, original nature of man, isolated individuals, economic cities, perfect states, and animal archetypes. One might add that all analogies belong in this category, e.g. God as the father of men, human behavior as if dictated by egoism, the State as an organism. Correlational symbols may also have this conjectural character if they refer to structures not clearly known to exist. For example, we often suppose a causal relation between occurrents which are not clearly known to be so related, and we frequently suppose a simple functional relation between two quantitatively determined occurrents when experience reveals only a highly complex relation.

It is the task of this chapter to determine the part which such hypothetical and constructional symbols play in science. But the problem reduces to two more specific problems: (1) Where do such symbols come from, i.e. how is their meaning determined? (2) For what purposes are they devised, i.e. what is their function? Clearly the two problems are interwoven. Since the derivation of symbols is an inventive operation, function seems

[1] Whitehead, *Concept of Nature*, p. 45.

to be logically more basic than origin. But since the effectiveness of a symbol in fulfilling its function is determined by the character of its operational derivation, origin seems to be logically more basic than function. I shall try to consider these two problems in the order given. Let us call all constructional and hypothetical symbols *suppositional symbols*. Then our problem here is to determine, first, the occurrential foundation and the operational route for the derivation of such symbols, and, secondly, the function which they fulfil in science.

Origin of Suppositional Symbols

There are apparently two ways in which this problem may be solved. One may say that suppositional symbols are *directly derived* from occurrents which are *obscurely recognized*, or *indirectly derived* from occurrents which are *clearly recognized*. According to the former solution one is obliged to say that they are *discovered*; according to the latter that they are *invented*. But in the former case they are not pure discoveries, for they are only obscurely recognized to be present; since they are indefinite as to content we can be sure only that we have discovered something,—the character of that thing still to be determined. So also in the latter case they are not pure inventions, for they are derived from their sources by operations which are describable and which are based upon certain features of the general structural system in which they participate. Thus we discover that the system makes certain demands beyond what is directly revealed, and in following out these demands we anticipate what experience is to reveal. Let us examine each of these methods in greater detail.

According to the former, a suppositional symbol is any symbol which is devised for the purpose of portraying an occurrent which is hinted at in a given situation without being clearly revealed. It is characteristic of the high abstractions and the high concretions of the world of occurrents that they are revealed to us in this obscure manner. Abstractions tend to become confused with associated abstractions and with the concretions in which they are found. For example, as we saw in chapter IV, it is not easy to distinguish between the pitch, timbre, and loudness of a tone, and for the untrained ear the tone is simply the undifferentiated complex of these abstractions. Abstractions may be more readily distinguished when they enter into awareness by independent sensory routes, as, for example, the taste and color of an orange. But even here the higher levels of abstraction are easily confused. So also high concretions are obscure because we tend to identify them with one or more of their less concrete elements; a highly concrete occurrent, such as a man, is identified by a few of his aspects, e.g. his ability to walk upright, to speak, and certain of his facial features. Hence there is a sense, perhaps, in which occurrents may be said to be given without being clearly given, or to be given only to those minds whose training has been such as to permit them to discern the occurrents in their larger structural situation. If this is true then the process of formulating symbols descriptive of these occurrents may be looked upon simply as an exploratory operation which turns the attention to the proper locus within the realm of occurrents, sets into functioning the operations of generalized characterization, and results in the direct derivation of symbols representative of these occurrents. This means, also, that in addition to the occurrents which are obscurely given in the situation, there are also given the structural relations which connect the obvious occurrents with the obscure ones. Since the relations can be no more clearly given than the terms, one must conclude that the structural relations are also obscurely given. Hence to say that the obscurely given occurrents are discovered puts something of a strain upon that word. Certainly they will be unconscious discoveries, for there is no distinct awareness of the relation; and the obscurity with which they are given leads us to think of them as inventions; finally, even if there is no distinct awareness, *for the discoverer himself*, of the dependence of the discovered entities upon their situation, if he is to convey his awareness to another individual he will be obliged to show how to get them from some obvious situation—and this makes them essentially inventive.

But there is a more significant inadequacy in this solution of the problem. If both the occurrents discovered and the route of discovery are indefinite and only obscurely recognized, how is it possible to say anything determinate and certain with reference to the character of the resultant entity? We cannot be sure *that* it exists at all; hence we cannot give it a forceful existential claim. And we cannot be sure *what* it is; hence to assert anything about it is to assert a conjectural proposition.[1] And there is the serious danger that what we do assert about it is simply the unjustified attribution to it of some feature possessed by the more obvious situation in which it is found, or, what is worse, the free and uncontrolled addition to it of fanciful, imaginary, and other highly conjectural properties. So long as the referent of the symbol is not given clearly we have no instrument by which we may remove the obscurity from the symbol. Hence through this method suppositional symbols become obscure and ambiguous, and tend to take on content not justified on any empirical grounds.

According to the alternative method all suppositional symbols have their foundations in occurrents, but are derived from them by elaborate rather than simple operational routes. Just as one derives simple characterizing symbols by direct generalization, so one derives suppositional symbols by such indirect routes as analogy, abstraction, negation, and idealization. This solution gives the suppositional symbol its proper occurrential foundation, and indicates how it loses something of its existential claim. But it leaves unexplained the problem of just how the content of the symbol is determined by the content of the origin. How is it possible to derive a *symbol of one content* from an *occurrent of another content*, and yet have the former depend on the latter?

I am convinced, however, that the method of operational derivation is incomparably more satisfactory than the method of the direct inspection of obscure occurrents. And only a slight modification of the method as formulated above is necessary. One may derive suppositional symbols from occurrents if one supposes that the derivation is complex. It seems possible, therefore, to describe the operational derivation of suppositional symbols as a two-fold movement consisting of (*a*) a direct inventive operation of generalized characterization performed upon a clearly given occurrent and resulting in a symbol representative of that occurrent, and (*b*) a direct inventive

[1] Cf. Hobson, *Domain of Natural Science*, pp. 32–3.

operation based upon a correlational symbol for a typical structural relation between occurrents, performed upon the symbol obtained in operaton (a) and resulting in the suppositional symbol. This enables one to see that every suppositional symbol arises by an indirect route from occurrent to symbol and symbol to suppositional symbol, and has its content determined through this operational derivation. The character of the second operation may be such as to obliterate more or less completely the resemblance between the suppositional symbol and its descriptive foundation. This is due to the two facts, to which we shall refer later, that operations may be largely indeterminate and negational in character, and that the content of the suppositional symbol may be increased after its original determination by all sorts of loose analogies. Thus molecules in motion do not seem to be at all the same sort of thing as heat; yet is seems clear that the notion of molecules in motion arose, originally and in its highly indeterminate form, out of the descriptive concept of heat. Hence we knew generally what molecules must do before we knew specifically what molecules were.

The fact of the indirect derivation of suppositional symbols permits the attribution of meaning to them in the absence of their direct referents. If the operational routes are properly defined, and if the occurrents which originate the operational routes are clearly given, the resulting definition of the symbols for the obscure occurrents may be made as precise as the determinateness of the operation permits. Since the occurrents themselves cannot be given clearly, we substitute for them their symbols. But since symbols are flexible and may be defined as we please to define them, subject only to the demands of communicability and of that which is to be communicated, we define the symbols by means of their structural relations to other symbols which are themselves directly exemplified. This is equivalent to defining the obscure occurrents in terms of those occurrents with which they are structurally related. But since both the occurrents and the structural relations are given obscurely we resort to symbols as substitutes for them.

It follows that the significant content of a suppositional symbol is structural. In order to make this clear let us suggest a distinction between three types of symbols. This distinction is not absolute, as can be clearly seen. But it affords a fruitful approach, and throws much light on the problem of explanation. Symbols may be divided from the point of view of unique and structural content into three classes.

(1) *Symbols whose meaning resides essentially in unique content and only incidentally in structural content.* Examples of such symbols can be found in the so-called psychological simples of experience, colors, shapes, sounds, tastes, etc. The contents of these symbols are determined directly by the occurrents from which they have been derived and they are verified in terms of such occurrents. They have very few structural relations, and such as they have are not considered essential to their contents; hence one does not derive their meanings from other more obviously given occurrents with which they are structurally related, nor does one verify them by finding them in anticipated structural relations with more clearly given occurrents. Their meanings are not determined through their situations but directly, by the occurrents themselves. The unique contents of such symbols seem to retain in an immediate sort of way the specific character of their referents.

(2) *Symbols whose meaning resides partly in unique content and partly in structural content.* Examples can be found in relative abstractions, e.g. number, order; in relative concretions, e.g. material objects, gold, a coin; in symbols for relatively extended spatial or temporal occurrents as well as relatively minute. For our purposes we are interested only in the symbols for the relatively high abstractions and concretions. The feature of such symbols is that the unique content cannot be clearly determined; hence we feel that both derivation and verification should be made in terms of structural relations. We feel that there *is* a unique content, i.e. there *is* that which has the structural relations, but the only way in which we can ascertain this content is to explore the structural relations. Thus we feel that a relative concretion is more than the mere associative union of its abstract elements, and we feel that a relative abstraction is more than a selected element of an associative complex. For example a coin is more than the mere associative union of a circle, a hard, a

weight, a color, etc.; it becomes in addition *that which* possesses a relation of inclusion to all of these occurrents. But this additional feature can be verified only in the existence of the associative complex. So also number is more than a selected element of an associative complex, say a pair of shoes; it becomes in addition *that which* possesses a relation of being-included-by to the pair of shoes. But this additional feature can be verified only by referring back to the associative complex. This is equivalent to saying that relatively high concretions and relatively high abstractions are not given so clearly as their elements and wholes respectively; hence symbols which designate these more obscure occurrents cannot have clearly definable unique contents.

(3) *Symbols whose meaning resides only incidentally or not at all in unique content and essentially in structural content.* Examples of these symbols are the very high abstractions and concretions, the very highly extended and minute, and all other symbols for obscurely given occurrents. Here both the derivation and the verification of the symbol are obtained through operational routes which are revelations of structural features of the given. The obscurity of the unique content has become so pronounced that we feel obliged to give to the symbols a suppositional status, i.e. we call them suppositional symbols. All unverified hypotheses and most partially verified hypotheses come under this category; we verify them not directly but in terms of their capacity *to explain*. To explain, as we shall see, is to reveal important structural relations. Thus we verify high abstractions by discovering situations from which they might be obtained by an operation of abstraction, and we verify high concretions by discovering situations from which they might be obtained by an operation of logical analysis. But we can never be sure whether the high abstractions and the high concretions themselves are given in the situations, for they seem to have only such content as is contained in the affirmation of their structural connections.

The content and existential claim of suppositional symbols will then depend upon two things: (*a*) whether or not the operational route is itself the expression of an actual structural relation, and (*b*) the degree of determinacy in the operational route. We discussed (*b*) in the preceding chapter with regard to some of the most typical of the correlational symbols. But we must now examine (*a*). Expressed crudely this principle states that suppositional symbols must be related to a system of directly descriptive symbols in a way which is characteristic of that system. We may be clearly aware of occurrents without being clearly aware of the structural relations of these occurrents to other more obscurely given occurrents. If these relations were given indubitably there would be no question as to the existential claim of the suppositional symbols. But since they are not certainly given we must infer them. What is the basis for such an inference? Apparently nothing more than the general uniformity of nature. Since most occurrents have more concrete and more abstract associates, more extensive occurrents which include them and less extensive occurrents which they include, repeated antecedents (i.e. causes, adjacent spatial occurrents), etc., we may infer with a relatively high degree of probability that any given occurrent possesses such structurally related elements. Since most symbols have implicates, implicans, compatibles, co-implicates, non-implicates, incompatibles, etc., we may infer with a relatively high degree of probability that any given symbol possesses such intensional content. The structural features of nature are probably more pervasive than its contentual features; i.e. there is a greater constancy in the way in which things are related than in what things are. Hence there is a greater probability that an occurrent will have, say, an abstract associate, than that this associate will be of a specific kind. But there will always be an element of probability rather than certainty in the inference as to the fact of a structural relation. Though we know that there is a general structural uniformity in nature we do not know that this specific situation possesses the structural relation in question.

We may summarize our discussion of the derivation of suppositional symbols as follows: Suppositional symbols are symbols for obscurely given occurrents. In order to avoid attributing the obscurity of the occurrent to the symbol for the occurrent, we define the symbol through an operational route. This is based upon a double operation which involves, first, the formation of symbols for clearly given occurrents, and, second, the derivation from these symbols through operations which are expressive of the typical structure of the system of symbols and which are relatively highly determinate, of other symbols which are called " suppositional " and are thus defined through their operational origin. This enables us to talk intelligently about obscure occurrents in advance of clear awareness of them, and in advance of any conviction that they even exist at all.

Function of Suppositional Symbols

But suppositional symbols cannot be understood if we consider them merely from the point of view of their origins. They play a role which is determined also by function. Simply stated, suppositional symbols aim to *explain*. The more detailed analysis of the problem of explanation will be postponed until chapter XIII, but we must raise certain preliminary considerations here.

In the broadest sense of the word explanation always involves the ascertainment of relations. In this sense to explain an occurrent or a symbol is to point out another occurrent or symbol with which it is related. This form of explanation does not exclude description, i.e., the designation of qualities, for to say that a thing has qualities is to say that it is associated with the occurrents exemplifying those qualities. But although explanation in this very general sense involves the revelation of structural relations, it is often employed to represent the discovery of only those relations which are highly determinative; that which is to be explained must be conceived as being *significantly* related to that in terms of which it is to be explained. We do not feel that we can explain an occurrent or a symbol unless we can increase its content by *derivation* from that to which it is related. The value of every explanation is that, however much we may know of the entity to be explained, we can learn something further about it through a specific structural connection. For this reason it seems somewhat misleading to say that explanation is always in terms of that which is more familiar.[1] We must always know something about the explaining entity which we do not know about the entity to be explained,

[1] As is suggested by Campbell, *What is Science?*, p. 77.

but by and large we may know much more about the latter than about the former. For example, we must know something about molecules which we do not know about heat, but in general we know much more about heat than we do about molecular motion.

This results in an important distinction between two different kinds of explanation.[1] On the one hand, we may explain that which is obscurely given in terms of that which is clearly given; but, on the other, we may explain that which is clearly given in terms of that which is, by and large and in the final analysis, more obscurely given but which reveals certain features in terms of which we hope to make the clearly given still more clearly given. No matter how much we know about an occurrent we always learn more through a discovery of further structural relations; and although the fact that these structural relations involve relatively obscure occurrents somewhat weakens the explanatory value, with the progressive clarification of the obscure entities there occurs a parallel clarification of the clearly given entities.

The former type of explanation is best illustrated in the process of making clear the content of a symbol by designating the occurrent from which and the operational route through which it has been derived. In fact, the entire discussion of chapters VII, VIII and of the present chapter up to this point may be taken as a description of this kind of explanation. The process of deriving symbols from their occurrential foundations is a process of explaining the contents of these symbols in terms of things which are more clearly given. In general we know more about occurrents than about simple symbols, more about structures of occurrents than about correlational symbols, and more about the fact of occurrents having structural associates than about either these associates themselves or the suppositional symbols by which they are designated. Hence the denotative method is essentially a method of explanation; we explain when we point, even though that pointing may be through a complicated operational route. And we can explain symbols ultimately *only* through denotative gestures; hence to show how a symbol is

[1] Cf. Eddington, *Nature of the Physical World*, pp. 249–50.

derivable is the only method by which its content may finally be understood.

But our concern at present is with the other type of explanation, viz. the explanation of the clearly given in terms of the less clearly given. This is illustrated by all of that aspect of science which involves the explanation of indubitable data in terms of hypotheses, theories, fictions, and constructs. In the terminology thus far introduced, we explain in this sense when we show that symbols for clearly given occurrents have highly determinative relations to suppositional symbols. But in this case the determination passes in the reverse direction from that in which it does in the problem of the explanation of suppositional symbols from data. In that case we wished to determine the content of the symbol from the clearly given occurrent; in this case we wish to determine certain further features of the clearly given occurrent from the content of the suppositional symbol. Thus in the former case the symbol is inferrible from the situation, but in the latter case the situation is inferrible from the symbol. These formulations are illuminating but slightly inaccurate. Properly speaking, we infer the content of the suppositional symbol not from the situation but from the symbol which immediately and directly describes the situation; so also we infer from the content of the suppositional symbol not the situation itself but a symbol which purports to describe the situation immediately and directly. This precision must be introduced in order to avoid the misconception that symbols may be *inferred* from occurrents; inference is a formal process of deriving symbols from symbols, and the derivation of symbols from occurrents as well as the discovery of occurrents exemplifying symbols cannot be looked upon as inferential processes.

We may now raise the question as to how suppositional symbols can explain. It hardly seems likely that a suppositional symbol can be turned about and used to explain precisely the situation from which it was derived.[1] If we know about the symbol only what we have been able to derive from the

[1] Cf. Carmichael, *op. cit.*, p. 34. The author here distinguishes a logic of discovery from a logic of demonstration. Cf. Russell, *Mysticism and Logic*, p. 146: "Waves are in fact inferred from the colors, not vice versa."

situation we cannot expect to derive from the symbol anything novel about the situation. An explanation of this kind would be purely verbal. Examples of such pseudo-explanations can be found in all use of such concepts as potentials (e.g. potential energy), capacities (e.g. the dormitive capacity of opium), unconscious purposes (e.g. *élan vital*), tendencies, instincts, affinities (e.g. chemical), powers, urges, desires, ideals, aims, causal efficacies, etc., where the suppositional symbols are defined in each case simply in terms of what they are known to produce.[1] Other examples can be found in the explanation of complex occurrents in terms of their abstractions when we know of the abstractions only that they are elements of such complexes, and of abstractions in terms of concretions when we know of the concretions only that they are complexes of such abstractions; in the explanations of parts by wholes when we know of the wholes only what we have inferred from the parts, and of wholes by parts when we know of the parts only what we have inferred from the wholes; and in the explanation of the present in terms of the past and the future when both the past and the future are only what we have inferred from the present. There is reason to believe that contemporary science allows explanation to rest in this preliminary state more commonly than is recognized. Its only virtue is that it locates a problem somewhat more specifically; its danger is that it will be taken as a substitute for explanation.

It should be recognized, however, that a theory may receive a high development in this stage before it passes into that stage in which it becomes more adequate as an explanatory notion. One of the most famous examples of this is the relatively high determination of the character of the planet Neptune upon the basis of what Neptune was to be called upon to do, viz. to explain the perturbations of Uranus. Other examples are found in the determination of the chemical properties of the missing elements in the chemical table upon the basis of the properties

[1] The method of multiple hypotheses discussed by T. C. Chamberlin in *Science*, Feb. 7, 1890 is another example of such pseudo-explanation—at least if it is taken as the final explanation rather than merely as working method. Here the attempt is made to include within the explanatory notion all of the features of the situation, even though it involves making the former as complex as the latter.

of adjacent elements, the determination of the character of missing organic forms by filling in the gaps between closely similar fossil remains, and the determination of the character of the ether from a knowledge of the nature of the energy it is obliged to transmit. In fact, one may say that this method becomes increasingly important in any science when the introduction of suppositional symbols has been retarded. For if there has been an extensive accumulation of data before the advent of any theory, the theory may then to a great extent be "read out of" the data. Usually, however, the data are so meagre at this point that one can give the suppositional symbol only a very limited content.

Constructional and Hypothetical Methods

The proper understanding of this double function of explanation lies in the introduction of a very important distinction whose first formulation is usually attributed to Rankine,[1] but whose reiteration has been a feature of most recent philosophy of science.[2] The distinction is usually formulated in terms of the "abstractive" and the "hypothetical" methods. The terminology is unfortunate since the former if properly understood is not usually abstractive at all in the ordinary sense of the term. I shall therefore take the liberty of substituting the word "constructional" for the word "abstractive." The discussion which is to follow will justify this choice of words. The distinction at hand is, then, that between the *constructional*

[1] W. J. M. Rankine, *Miscellaneous Papers* (London, 1881), p. 210.
[2] E. Cassirer, *Substance and Function* (Chicago, 1923), pp. 193, *et seq*; Dingle, *Science and Human Experience*, chap. III; Hobson, *Domain of Natural Sience*, p. 132; Meyerson, *De l'explication dans les sciences* (Paris, 1927), p. 59; Bavink, *The Natural Sciences*, p. 34. Most of the dispute over Newton's "*hypotheses non fingo*" centers about the question as to whether he meant hypotheses in the ordinary sense, or constructs, i.e. abstractions. . . . Vaihinger's distinction between *fictions* and *hypotheses* (*The Philosophy of 'As if'* pp. 85–90) is not precisely the distinction here suggested, since for Vaihinger all fictions exhibit a "deviation" from reality, i.e. they contradict reality. But constructs cannot contradict reality though they may deviate from reality in the sense that they differ from something which obviously exists. To characterize all deviations from reality as fictions is to make all suppositional symbols (including hypotheses) fictions. For they all attempt to characterize occurrents which are different from those lying in the realm of the clearly given. Cf. also Ritchie, *Scientific Method*, pp. 158–9.

and the *hypothetical* methods for the derivation of suppositional symbols.

If we wish to elucidate this distinction by means of material already available we may say that suppositional symbols obtained through the constructional route must be explained in terms of their origin but symbols obtained through the hypothetical route must explain their origin. The key to the difference lies in the recognition that *the hypothetical method is a technique for the enlargement of the content of suppositional symbols which have already been derived by the constructional method*. In the latter method we are entitled to give to the suppositional symbol only such content as is derivable from the situation when the specific operation is performed; three of these methods are, as we shall see, the associative, the abstractive, and the concretive. But in the hypothetical method we start with the suppositional symbol as thus defined and by means of certain operations *not essentially inferential in character*, i.e. *not highly determinative*, we increase its content to such an extent as to permit inference from it to the situation from which it was originally derived and which we are now trying to explain. The question as to whether the hypothetical method is defined as beginning with the results of the constructional method or as including the constructional method as part of itself is perhaps relatively unimportant. Probably the latter interpretation is more in accord with the use of the word in scientific literature, and we shall therefore adopt this meaning. Then the hypothetical method is a two-fold method for the inferential derivation of suppositional symbols from occurrents through operational routes (constructional method), and for the non-inferential increase of the contents of such symbols through other operations designed to permit inferences from the suppositional symbol as thus increased to the given situation. The character of these non-inferential operations must be discussed in greater detail later; their common feature is the more or less free addition of content without strict regard for justification. No need is felt for justification since the tentative character of the suppositional symbol is clearly recognized, and any inferences which we then make from the suppositional symbol to the situation are presumably capable of being empirically verified.

The difference between the methods can be illustrated by the mathematical symbol " i " and the physical symbol " molecule." In both cases we start with a realm of relatively clearly given occurrents. In the former it is the body of real numbers and the structural relations expressive of their operational derivations from one another. In the latter it is that complex of phenomena including the three states of matter and their transformations into one another, the properties of compounds with relations to the properties of their elements, relation between pressure and volume of gases, expansive character of objects under increases of heat, vapor tension of liquids, solubility, etc. From the complex of numbers i is defined operationally, i.e. the symbol i is defined as what one gets when he extracts the squareroot of negative one. But what does i become as a result of this operational derivation? Since it is a constructional symbol, it possesses only such content as is permitted by inference from the character of the mathematical system as a whole. If every operation upon a number produces a number, then i is a number and possesses all of the properties and structural relations of numbers. There seems to be no important reason for concluding that we could learn something else about numbers in general if we only know something else about i. But if we should increase i through the hypothetical method and then succeed in deducing an important new propostition about numbers, we should not rest satisfied until we had shown how this additional property could be derived through the constructional method. The highly integrative character of mathematical systems usually permits this to be done; hence suppositional symbols in mathematics are almost entirely constructional in character.

On the other hand, although molecules were constructionally derived, they soon became hypothetical. The notion arose, and probably received its first content as distinct from the notion of atoms, through Gay-Lussac's discovery that gases always combine in volumes that bear simple ratios to each other[1]; molecules were then definable simply as the property

[1] W. C. D. Dampier-Whetham, *History of Science* (Cambridge, 1930), p. 229.

of combining in definite proportions. Later on, through the work of Count Rumford, Joule, and Clausius the identification of heat and energy made possible the formulation of the kinetic theory of heat; molecules then became definable as those things which in rapid motion exhibit the property of heat. Both of these seem to be constructional definitions. But the way is now prepared for the enlargement of the content of the symbol through the hypothetical route. The first step is taken with the identification of molecules-as-the-property-of-combining-in-definite-proportions with molecules-as-equivalent-of-heat; this is the result of the hypothetical method rather than the constructional for by no inferential process can we determine that one suppositional symbol explains both phenomena. By the assumption of perfect elasticity on the part of the molecules it becomes possible to deduce Boyle's law, Charles' law, and the law for the rate of diffusion of gases through porous partitions.[1] Although some of these principles were known before the formulation of the molecular theory, they were explained only by the hypothetical enlargement of the concept of molecules. So also by attributing to the molecules certain further properties it became possible to explain evaporation of liquids and compression of gases, solidification of liquids, vapor tension, and even some deviations from the gas law. Thus although molecules arose originally out of, and were therefore defined in terms of, certain situations, their significant content was not determined by inference from these situations but by loose analogies, guesses, imaginative experiments, and the like. They began as constructs, and they became hypotheses. So long as they are constructs they do not explain anything. They are to be understood simply as " entities which exhibit the properties revealed in experience." But if we know of the entities only that they reveal certain properties, we cannot explain anything in terms of them. Hence we must presume that if they exhibit the known properties they probably also exhibit certain other properties, from which it may be possible to explain, i.e. to derive the data. At this stage the suppositional symbol has become hypothetical.

Let us endeavor to see more clearly the essential features of these two types of method.

[1] *Ibid.*, p. 249.

Constructional Method

The constructional method is that part of the hypothetical method which is concerned with definition of the symbols for obscure occurrents in terms of their origin; but it is concerned with this definition only to the extent of giving the suppositional symbols such content as they may be *inferred* to have upon the basis of the origin and the operational route, and not such content as they may be *supposed* to have if they are going to explain. This is probably what led Rankine and others to speak of it as the abstractive method. But it cannot be identified with the abstractive method, for abstraction is only one mode of inference to obscure occurrents. Constructs do not explain, for we determine their contents from data, and we cannot therefore infer anything new from them to data. All of the pseudo-explanations listed earlier in the chapter, e.g. potentials, capacities, unconscious purposes, are cases of constructs rather than hypotheses. Hence as explanations they must be purely verbal.

The most important of the constructional methods are three in number: the *associative method*, the *abstractive method*, and *concretive method*.

(1) *The associative method.* The associative method is that operational route which is expressed symbolically in the correlational symbol of co-implication, and which is based upon the fact of repeated associative complexes. We employ the associative method when we start from a symbol for a clearly given occurrent and pass to a suppositional symbol related to it by co-implication, and then determine the content of the suppositional symbol through this operation. This is the method which we always employ when we infer certain more obscure features of an object from certain other features more clearly given, e.g. the taste of an apple from its smell, the weight of a stone from its texture, or the smoothness of a table from its gloss. Many factors contribute in such situations as these to destroy very quickly both the obscurity and the hypothetical character of the suppositional symbol. In the first place the operation is very highly determinate both as to content and as to existential claim. All of the structural relations of the clearly

given occurrent may be transferred to the suppositional symbol; this is founded upon the basic principle that if a symbol participates in an associative complex, it participates in any occurrent contained in that complex. This means that by a careful examination of the clearly given element we may determine in advance many features of the obscurely given element, e.g. by a careful inspection of the character of the gloss we can determine the character of the smoothness. The high determination as to existence lies in the fact that the obscure occurrent must be found (if the symbolic correlation is not itself erroneous) and it must be found in precisely the given situation. In the second place the foundation of the operation is not merely the general structural uniformity of nature but the specific fact of the repeated associative complexes in the past. Thus we do not rest content with founding our inference on the mere fact that most occurrents have associates, but on our generalized experience to the effect that occurrents of the specific kind in question have associates of another specific kind. In the third place we do not define our symbol *wholly* in terms of its operational derivation, for we know from previous experience what the symbol means. For example we do not define smoothness as that which one experiences when he touches a glossy surface. We know in advance what the suppositional symbol means, and we know in advance that this meaning is essentially a matter of unique rather than structural content. Finally the obscurity in the associated occurrent is not inherent since we may usually eliminate it by a simple physical operation, such as focusing the eyes, turning the head, extending the arm, biting, sniffing, moving fingers over the surface, pinching, or pressing. The suppositional symbol then loses completely its conjectural character.

This fact of the ready transition of such suppositional symbols into symbols for clearly given occurrents somewhat limits their use in the scientific method. They are only temporary, and merely instruments of exploration. They tell us nothing about the occurrents which are inherently obscure, e.g. high concretions and abstractions, highly extended and highly minute, and past and future. For this reason the abstractive and the concretive methods are more important.

(2) *The abstractive method.* The abstractive and the concretive methods are both operational routes which are expressed symbolically in the correlational symbol of implication, and which are based upon the fact of asymmetrical repeated associative complexes. The difference in the two methods lies in the direction of the operation. The abstractive method determines the content of a suppositional symbol from the content of a more directly descriptive symbol which is its implicans. Thus the abstractive method is the route for the definition of abstractions in terms of concretions. If highly abstract occurrents are inherently obscure then an empirical theory of meaning must show how that obscurity may be removed. This cannot be done through other equally high abstractions, nor through even higher abstractions; it must therefore be done through greater concretions. This seems to me to be in essence what we do when we elucidate abstractions, i.e. we point to their more concrete associates. We explain what shape is by indicating a triangle, what tone is by sounding a note on the piano, what number is by pointing to a pair. But we try to indicate that shape, tone, number, etc. are not these but *abstractions* from them. Thus we invite the listener to define the abstraction in terms of the concretion so far as the structural features of a concrete may ever be transferred to an abstract. The principles of this transfer we discussed in the last chapter. It must be noted here that the abstraction is not from the occurrent to the symbol; there is probably some sense in which all occurrents are concrete and all symbols are abstract, but this distinction between these two terms has not been employed at any time in this discussion. Hence when I say that we define an abstraction in terms of a concretion, I mean that we define a symbol for an abstract occurrent in terms of a symbol for a concrete occurrent, and then we define the symbol for the concrete occurrent by the denotative method. Thus there is always an indirect route from the occurrent itself to the suppositional symbol.

The fact that abstractions are less specific than their concretions leads one to think of the abstractive operation as being essentially negative in character.[1] Hence one may apparently form abstractions from concretions by a mere denial of certain of their aspects. It seems clear to me that many of the so-called fictions and constructs of science are defined in this purely negational way. A perfect lever is a lever *without* the imperfections of experience, perfect motion is motion *without* friction, a point is position *without* extension, a perfect state is a state *without* internal discord. But mere negation is an indeterminate operation; to know that two occurrents differ in one respect does not tell us anything about them in other respects. Hence suppositional symbols which are defined in this way are essentially obscure, for one never knows precisely what meaning to attribute to the resulting negation.

But if one understands the character of the abstractive operation he will see that it is not purely negational. He will see that all of the implicans of the concrete must be also implicans of the abstract and all of the compatibles of the concrete must be compatibles of the abstract. Other less important relations may be transferred. Furthermore he will see that to form a negation is not necessarily to form an abstraction; we often lose properties in passing from abstractions to concretions as well as the reverse.

However, it is probably true that most of the constructs of science can be defined through the abstractive route, if the nature of that route is properly understood. The literature of recent philosophy of science abounds in such attempts, e.g., Whitehead's definition of points and instants through extensive abstraction,[2] Dedekind's definition of the irrationals,[3] Cantor's definition of the infinite,[4] Russell's definition of number,[5] Broad's definition of matter in terms of sensa,[6] Lenzen's definition of the concept of the physical order from the concepts of body and space-time coincidences,[7] Stace's definition of the external

---

[1] The point is well brought out by Cassirer, *op. cit.*, chap. I.
[2] *Concept of Nature*, Chap. IV.
[3] R. Dedekind, *Essays on the Theory of Numbers* (Chicago, 1909), esp. p. 15.
[4] G. Cantor, *Contributions to the Founding of the Theory of Transfinite Numbers* (Chicago, 1915), esp. p. 104.
[5] *Principles of Mathematics* (Cambridge, 1903), chap. XI.
[6] *Scientific Thought*, Part II.
[7] *Psysical Theory*, esp. p. 48.

world in terms of the stuff of the world of solitary mind,[1] Carnap's definition of *physische, fremdpsychische* and *geistige Gegenstände* in terms of *eigenpsychische Gegenstände*[2] and so on. To say that all such entities are logical constructions is harmless so long as one does not then go on to deny them metaphysical status. Perfect levers and triangles, mathematical points, infinities, utopias, completely isolated human beings, etc., are logical entities, for they are derived and defined from operational routes which are expressive of the structure of existence. But to make them abstractions does not prevent them from existing, though we must not expect them to exist in the same obvious way that their concretions do. But this does not mean that they must be put into an ideal realm, or a realm of fictions, or a realm of subsistence, or a realm of mathematical entities—supposing that each of these realms exhibits features essentially different from that of the empirically given. On the contrary, they are part of the metaphysically given, though they are given less clearly, and they must be defined and located through operational routes. For example, when we define and locate infinity we direct the attention of the individual to a series of classes arranged in order of size ; then we point out that infinity is the abstraction from the ordinal relation of constant increase ; finally we define infinity in terms of this descriptive symbol and the relatively highly determinate operation of abstraction.

One of the most important applications of the abstractive method is to the field of the definition of the so-called basic concepts of the sciences. The operational definition of these concepts constitutes the main task of that part of the philosophy of science which was called in chapter I the metaphysics as opposed to the logic of science. Its problems are indicated by the following : How is " number " as used by the mathematician, i.e. as including negatives, fractions, irrationals, imaginaries, etc. derivable from " number " as it obviously exists, e.g. as a property of a pair or of a triad ? How is " Newtonian time " as a single, absolute, all-embracing, empty, homogeneous entity derivable from " empirical time " as many, relative, independent,

---

[1] *Theory of Knowledge and Existence*, chap. VI.
[2] *Der logische Aufbau der Welt*, Abschn. III.

occupied, heterogeneous entities? How is the "space" of geometry derivable from the "space" of direct experience? How is the "matter" of the physicist derived from the "matter" of the common sense individual? All of these derivational routes are essentially abstractive; the attempt is made to define the "scientific" notion in such a way as to make the "common sense" or "empirical" notion a specific kind or type of which the concept as it functions in science is a generalization. It is the task of the philosophy of science to examine these operational derivations, and to estimate the accuracy with which a concept is defined through this route.

(3) *The concretive method.* The concretive method is also based upon the correlational symbol of implication but in this case the operation is from the implicate to the implicans. The concretive method is the route for the definition of concretions (complexes) in terms of abstractions (simples). This is the method which is ordinarily called *definition* and there is no objection to limiting the word to this situation. But it must not be forgotten that one can also "define" abstractions in terms of concretions. Thus one may say that the abstractive and the concretive methods are both routes for the determination of symbols from symbols and are hence "definitions" in the general sense, but the determination of concretions from abstractions is "definition" in the special sense. The assumption of this method is that highly concrete occurrents are inherently obscure; consequently an empirical theory of meaning must show how that obscurity can be removed. But this cannot be done through equally high concretions nor through higher concretions; it must therefore be done through greater abstractions. This is in essence what we do when we analyze (in this specific sense) complex occurrents. We explain what a table is by indicating its shape, stuff, color, weight, etc.; and we explain what a man is by pointing out his animal features, his material features, his rationality, etc. But again we try to indicate that the table and the man are not simply these aspects but *concretions* formed from them. Thus we invite the listener to define the concretion in terms of the abstraction so far as the structural features of abstractions may ever be transferred to concretions. We

discussed the principles of this transfer also in the last chapter. The most important features are two in number: (a) The operation to concretions usually starts not in an isolated occurrent but in an associative complex; the concretion becomes precisely the associative complex, i.e. the whole which contains the associated elements. Thus the ether was at one time defined as that which is both liquid and solid; light has recently been described as both wave-like and corpuscular; and in general any scientific hypothesis takes on the totality of attributes which experimental situations demand that it possess. This is an example of the concretive method, for from an awareness of occurrents one passes to the awareness of the symbol for the associative complex of the occurrents; this occurrent then becomes the concrete occurrent of which the given occurrents are abstractions. (b) The concretive method permits the transfer of implicates but not of implicans. Thus one may say, crudely, that the attributes of the elements are the attributes of the complex, though the complex may possess implicates which the elements do not possess. This is probably what is meant by saying that wholes are emergents and exhibit features which are the result of organization. Russell has tried to avoid the dangers of unjustified transference of properties to the complex by defining a thing as the class of its perspectives rather than as that which possesses the perspectives,[1] by defining class extensionally rather than intensionally,[2] and in general by preferring any association of elements to a whole which contains them. I am unable to see that this is more than a verbal solution; for the fact of an association determines immediately an occurrent which is the complex, an occurrent which is different from either of the elements and may therefore possess unique structural relations.

These are some of the most important of the operations for the derivation of suppositional symbols. Other operations, of which only a few can be mentioned, are (1) causal inferences, through which suppositional symbols designating causes are derived from and determined by symbols for clearly given

[1] *Our Knowledge of the External World*, p. 89.
[2] *Principles of Mathematics*, p. 69.

effects, or suppositional symbols for effects are derived from and determined by symbols for clearly given causes ; (2) analytic and synthetic operations by which suppositional symbols for extended wholes and minute parts are derived from and determined by symbols for clearly given parts and wholes ; and (3) spatial operations by which symbols for distant occurrents are defined. The general principles involved in all of these operational routes are the same, though the nature of the specific method is determined in each case by the correlational structure upon which it is based.

Hypothetical Method

Strictly speaking the hypothetical method includes the constructional method. Here, however, we shall presume the results of the constructional method and endeavor to point out what the significant additions of the hypothetical method are. We have already seen that the latter method aims to increase the content of the construct in such a way as to make it explanatory of the given. Constructs are relatively indeterminate as to content, hence the task of the hypothetical method is the progressive increase of this content with a view to the explanatory possibilities of the resulting symbol.

The example of the mathematical symbol "i" and the physical symbol "molecule" reveals two important features of the hypothetical method. In the first place it suggests the error involved in considering the hypothetical method as an act of essential mystery, arising without preparation and direction, yet having the remarkable property of producing a suppositional entity which explains the situation out of which it arose. There is undoubtedly something of mystery in this act ; it is variously characterized as a flash of intuition, an act of imaginative insight, a sudden illumination. The act seems to occur apparently without direction or control and under the greatest possible variety of conditions—the concentration of the laboratory, immersion in intense mental activity of an unrelated sort, or periods of relaxation such as a walk in the country. Yet the remarkable feature of all of these acts is that they produce explanatory notions having a direct relevance to the situation in question; the cry "Eureka, I have found it" is a necessary accompaniment of all acts of scientific discovery. This fact, together with other considerations which I have raised elsewhere,[1] seems to me to point to the fact that the act of scientific discovery is much less of a mystery than it is commonly supposed to be. I do not refer, of course, to the psychological factors which are responsible for the occurrence of the act at one time rather than at another; these aspects of the situation are still wrapped in relative obscurity.[2] But apart from these aspects the act seems to be mysterious only because we tend either to overlook or else to belittle the importance of the constructional method as part of the hypothetical method. Through the constructional method we give the suppositional symbol not such content as it must have to explain the situation, but such content as it must have if it is even to be compatible with the given situation. In the constructional method we determine the possible range of content within which the suppositional symbol must be further refined. This is determined by direct inspection of the situation and the operational derivation, and thus we see why it is necessary for the investigator to be thoroughly familiar with his data before he can expect the flash of insight to occur. Important discoveries by the untutored are very rare in science.[3] If we recognize that this constructional act is a prerequisite to the hypothetical act which involves an increase in the determination of the hypothetical symbol, we have explained away at least part of the mystery. For now instead of solving the problem of how an investigator can devise an explanatory notion from a situation we have only to show how from a suppositional symbol having a relatively indeterminate content which is derivable from a situation we may devise a more specific content which will explain that

[1] In an article "The Mystery of Scientific Discovery," *Philosophy of Scienc* Vol. I, pp. 224-236; cf. Carmichael, *op. cit.*, chap. I; cf. also Cohen's use of the term "disciplined imagination," *Reason and Nature*, p. 60.
[2] Cf. J. M. Montmasson, *Invention and the Unconscious* (New York, 1932); Th. Ribot, *Essay on Creative Imagination* (Chicago, 1906); G. Gore, *Art of Scientific Discovery* (London, 1878), part III; E. Rignano, *Psychology of Reasoning* (London, 1923), chap. VI; Graham Wallas, *Art of Thought* (New York, 1926), chap. IV; C. Spearman, *Creative Mind* (New York, 1931), chap. I–IV, VIII.
[3] Cf. Bavink, *The Natural Sciences*, p. 240.

situation. To be sure, the actual hypothetical methods may still be mysterious but they cannot be as mysterious as the total derivational act because we have already removed the obscurity in part of this act.

In the second place our example reveals the importance of the suppositional method. There is a tendency to suppose that this method by itself is of no value to science; we have already seen that if we turn about and endeavor to explain our situation through symbols derived directly from it through the suppositional method, our explanation becomes merely verbal. A symbol derived through this method alone is not sufficiently determinate. Putting it crudely, we may say that we know a few things that the suppositional symbol must *do*, but we do not really know what it *is*. It has no informational value, and we can extract nothing new from it. But such a symbol should not be condemned, for it represents the stage through which every hypothesis passes in its formulation. There is a time when every hypothesis possesses its content necessarily, simply because it was inferentially derived from a situation exhibiting certain features. This represents the permanent content of the hypothesis, and however the hypothesis may be increased as to content in the future this feature must remain[1]—supposing of course that the data themselves are not erroneous. Specific hypotheses may come and go, but the general content of the hypothesis remains. Hence the derivation of the hypothesis through the constructional method gives it a certain minimum of content which it must possess if it is to be compatible with the situation. This content cannot be taken away from the hypothesis, for if it is once compatible with known fact it must always be so. This is why many discarded theories in science still retain a certain working value. One of my colleagues who is a teacher of chemistry remarked at one time that it is still convenient on certain occasions to employ the caloric theory in the explanation of heat; if this is so it must be due to the fact that abandoned theories are never completely rejected but rather made over into

---

[1] Cf. Poincaré, *Foundations of Science*, p. 140; H. Weyl, *Mind and Nature* (Philadelphia, 1934), pp. 46–47; H. Feigl, *Theorie und Erfahrung in der Physik* (Karlsruhe, 1929), pp. 129–30.

new theories which retain the valuable features of the old in taking on the additional content of the new.

The essential feature of the hypothetical method can be seen to lie in its double claim. Supposing the results of the constructional method, the hypothetical method professes to be a technique by which we may derive symbols having both a more determinate content than the suppositional symbols from which we started and an existential claim which is a maximum as compared with the claim of these symbols. In other words through this method we hope to make the contents of the symbols more specific without losing any more than is necessary of the existential claim. That there will be some loss in existential claim is inevitable. The structural organization of occurrents implies a general principle of probability, viz. that an indeterminate entity is more likely to occur than a determinate one. This is based upon the nature of the concrete-abstract relation between occurrents, which is such that from a concrete occurrent we may conclude the existence of an associatied abstract occurrent but from an abstract occurrent we may not conclude the existence of a (specific) concrete occurrent. As we increase determination we lose existential claim; high abstractions are pervasive, high concretions are not. This is illustrated by the assertion that if there are inhabitants of Mars there is greater probability in favor of the proposition that they are merely living beings than that they are exactly like ourselves. It follows that if the hypothetical method is to insist upon the increase in determination, it cannot insist also upon the retention of the existential claim. As a result hypotheses have a relatively weak existential claim as compared with constructs. There is no harm in this, since the function of the verificatory processes is precisely that of increasing gradually the existential claim of the hypothesis. As a hypothesis is found to explain a wider and wider range of phenomena and to imply no propositions contrary to the known range of phenomena, its existential claim increases.

The specific nature of the hypothetical method may now be considered. Supposing that our suppositional symbol has already been given content through the constructional method, what are the techniques for increasing that content in such a way as to give the suppositional symbol explanatory value?

They will be of two kinds, which are determined by the double demand which we put upon explanatory notions. On the one hand we may be satisfied if our explanatory notion is such as to account for the *known data*; on the other hand we may insist that our explanatory notion be such as to permit the inference to symbols for *data not yet discovered* but presumably present in the situation. According to the former demand we are satisfied if our symbol merely explains, but according to the latter we require that our symbol have a certain working value in the direction of fostering research and advancing knowledge. Both of these methods are essentially hypothetical rather than constructional, for we do not infer the suppositional symbol from the situation but rather we conjecture as to what content the suppositional symbol might have which would permit us to infer the situation (as it is or as it is increased) from it. If this is to be called inference at all, it should be called *inductive* inference. For the present we shall do well not to characterize it as inference at all, though we shall see immediately that it exhibits at times a relatively high degree of determination.

Of these two hypothetical methods the former creates the distinct impression of cheating. Granting that the constructional method has been properly applied, we know what content our suppositional symbol must have if it is to be compatible with the situation. But how is this content to be made more determinate? By deliberately adding to the suppositional symbol such content as will permit the deduction of symbols describing the known situation. Hence one knows what the implicates of his hypothesis should be and then he sets about to construct an hypothesis which will be precisely of this character. It is not surprising, then, if the hypothesis *does* explain the situation, though it may appear so to one who does not know how the hypothesis was derived. When the magician takes a rabbit from a hat, he surprises the audience though he does not surprise himself, because he put it there. It is important to recognize that there are hypotheses of precisely this character. They are called *ad hoc* hypotheses.[1] Their essential feature lies in the fact that they explain the situation out of which they arose, but they do not go beyond it; they have no predictive value for further elements of the situation or for other situations of the same general kind. The hypotheses become more and more complex as new elements of the situation are discovered. Further, such hypotheses have no general explanatory value, and we should need as many as we have diverse situations to be explained. *Ad hoc* hypotheses are frequently employed in violation of the principle of parsimony; for example, if the actual outcome of an experiment is not in accord with the result anticipated by the hypothesis, an *ad hoc* hypothesis may be introduced under the form of an " interfering factor." Then all that we know of the hypothesis is that it is such as to explain the interference. There is some reason to believe that the action of catalytic agents in chemistry can be " explained " only in this way in our present state of knowledge. The remark frequently uttered by scientists to the effect that we know much about what electricity *does* but little about what it *is* leads one to believe that perhaps electricity is itself an *ad hoc* hypothesis.

The point is that every hypothesis passes through the *ad hoc* stage just as it passes through the construct stage. First we give our suppositional symbol only such content as is demanded by the situation; then we gradually increase it in such a way as to permit the deduction of features which we know the situation to possess and which themselves are the basis for the increase in the content of the suppositional symbol; finally we increase it in such a way as to permit the deduction of features which we do not know to be present in the situation but which we hope will be revealed by further exploratory processes. It is only at this stage that the hypothesis becomes full grown, for it now possesses those directive features which make it so valuable. But by becoming more determinate in character it has further lost its existential claim. Thus we speak of hypotheses as being conjectural, problematic, doubtful. The route of derivation has now become so intricate that the existential reference has

[1] Cf. J. Rueff, *From Physical to Social Science* (Baltimore, 1929), pp. 48–49. One gets the impression from Rueff that all hypotheses are of this character. Cf. also p. 23.

been very much weakened. But to compensate, the hypothesis has a relatively specific content; though we do not know whether it exists, we know pretty clearly what it means, whether it exists or not. We must now turn to a consideration of these processes for the increase of the content of the suppositional symbol.

Since these hypothetical methods are devised rather for increasing content than for retaining existential claim they will have a low inferential value. Hence they are essentially methods of suggestion, and will consequently be incapable of formalization. Most prominent among them will be crude analogies and contrasts, wide and sweeping generalizations, quantitative increases and decreases, imaginative experiments, and the like— all of which in their actual employment are little more than somewhat highly controlled guesses, and many of which turn out to be essentially bad guesses. But they serve their purpose in that they give to the suppositional symbols the necessary specificity which will permit the drawing of deductions. Three of these methods may be mentioned in passing, viz. analogy, imaginative experimentation, and generalization.

Analogical operations are based upon the fact of similarity. Similarity is a relation between symbols and between occurrents which is based upon the fundamental fact of spatio-temporal structure. Two occurrents are similar if the same symbol participates in their respective situations. But similarity is a matter of degree. With reference to symbols it is determined by the number of structural relations which the two symbols have in common. If two symbols are co-implicative they are completely similar as to structural contents, for all of the structural relations of the one may be transferred to the other. We seldom have such perfect similarity. The most common meaning of similarity is that in which two symbols have a common implicate; in such a case we say that the symbols represent species of a common genus. But there may be a diminishing similarity down to a minimum in which the two symbols have only one or two structural relations in common. It is clear that if two symbols have many relations in common there is greater reason for believing that they will possess a further relation than would be the case if they had but few. Of course the matter of the essentiality of the relation also enters in. But we may say that an analogical operation is one by which awareness passes from a symbol A which is known to imply C, to another symbol, B, which is also known to imply C, with the result that some of the further structural content of A (say, the fact that it is implied by D) is transferred to B. Then B in the original form would constitute a suppositional symbol derived only through the constructional method, but in the final form it would constitute a hypothesis to be subjected to verification in terms of its consequences. For a simple illustration we may take the following. Sound, A, which exhibited certain properties, C, was like light, B, which seemed to exhibit much the same properties. But sound was known to be explicable in terms of wave phenomena, D; hence it seemed reasonable to suppose that light could also be explained as a wave phenomenon. From this hypothesis certain consequences, later verified, were drawn. The closeness of the analogy determines the probability with which the hypothesis may be accepted.[1]

One very important type of analogical operation is that which might be called serial extension or passing to limits. This is a method which is employed in defining idealized or perfect entities where there is conceived to be a continuous gradation of objects organized according to the principle of increasing perfection. Such a series is finite but it indicates an extension beyond itself and to the definition of the entity found at the end of the series. Thus a perfect line is the limit of lines arranged according to the principle of decreasing thickness. So also perfect levers, gases, societies, etc., are defined. This is a method which is often employed as alternative to the abstractive method described earlier in the chapter. It produces entities with a much weaker existential claim since the principle of inference is analogy rather than abstraction. The sample upon which the analogy is based is that type of series which exhibits

[1] Bridgman (*Logic of Modern Physics*, pp. 40–42) suggests that as we approach the very small, certain discontinuities appear; hence nature is not everywhere alike. This simply means that analogical inference from the large to the small has a low probability. But employed as a method of hypothesis it involves no essential dangers.

not only an indefinite increase but also a limiting member, e.g. the series of regular polygons with increasing numbers of sides with the circle as a limit. A series of lines exhibits the same property of continuous increase as the polygons; hence it is reasonable to suppose that there is also a limiting member which is a perfect line. The weakness of the analogy reduces the existential claim. Hence there is more reason to believe that there is a perfect line if this is defined as the property of continuous increase than if it is defined as the limit of that increase as the number of lines approaches infinity.

There is a second type of hypothetical method which may be mentioned. This may be called the imaginative route, since it is based upon thinking in icons rather than in characterizing symbols. It is another method for increasing the content of a suppositional symbol when it has been derived by the constructional method. The principle is that the pictorial representative of an occurrent is always more definite and specific than its word-symbol. Hence one has only to take the characterizing symbol as defined through the constructional route and substitute for it an image or picture of the occurrent thus referred to. This involves a visualization of the construct. Such an attempt at concrete imagery will always involve an increase in specificity since we dislike vague, indefinite images. Here there is no principle whatsoever guiding the transformation. This filling in of imaginative content will depend upon a number of individual, psychological factors such as the characteristic mode of imagery of the individual concerned, the range and content of his experience, and his training in the formation of images. It is probably true that for certain individuals much thinking is carried on in this way, even that which is concerned with making further inferences from the hypotheses, i.e. making predictions. One may picture a situation in his mind and then deliberately set out to modify the image content in certain ways for the purpose of seeing what further modifications may be observed to follow. These modifications may then be presumed to be what actual experience would reveal. This is probably what Rignano[1] means by his thought-experiment. Such a mode of thinking is precarious because there are no principles for the estimation of its adequacy and because of the impossibility of retaining all of the features of the image situation in mind. However, the method is often employed, and may even be the foundation for important scientific discoveries. We shall return to it in chapter XI.

A third type of hypothetical method should be mentioned in closing, since it indicates that operational elaborations may be performed upon correlational symbols was well as non-correlational symbols. This is generalization as an operational route for making universal a correlation which is not known to be more than occasional. If the datum entitles us to symbolize a situation as "A is compatible with B," we are unable to explain this situation until we are able to derive the proposition from some other proposition. We then formulate the hypothesis that A implies B. There is no strictly inferential justification for this increase in content, though attempts have been made through " inductive syllogisms " and the like to formulate such principles, and logic books have discussed some of its most characteristic errors, e.g., composition, converse accident, and *post hoc ergo propter hoc*. There is clearly no harm in these attempts since if the process can be formalized it is less likely to be erroneous. But having once taken place, the operation finds its justification in its results ; if we continue to find further associative complexes of the same form and none of a contradictory form, we are entitled to assert that the hypothetical increase was justified.

The foregoing chapter has attempted to reveal the processes involved in the formation of suppositional symbols. We have seen, briefly, that such symbols are derived not from occurrents but from symbols which refer to clearly given occurrents. From such symbols suppositional symbols take on their contents through a complex operational route which involves first a constructional method and then a hypothetical method. The former gives the symbol such content as we may infer it to possess upon the basis of the content of the descriptive symbol and the

[1] *Psychology of Reasoning*, chap. IV ; cf. also Mach, *Erkenntnis und Irrtum* (2nd ed.; Leipzig, 1906), p. 41.

nature of the operational route. But a symbol in such a form has no explanatory value. Hence we pass to hypothetical methods which endeavor to make the content of the suppositional symbol more specific by the relatively unguided addition of features. By means of the symbol as thus constituted the original situation may be explained. We are now ready to turn to a consideration of the general claims of knowledge and the possibility of systems of knowledge of varying degrees of adequacy.

CHAPTER X
THE DEVELOPMENT OF KNOWLEDGE

The task of science, we have seen, is that of deriving from the realm of occurrents through operational routes a body of symbols which is then presumed to be representative of that realm. Knowledge arises only when, associated with the awareness of the realm and the awareness of the symbols, there is the awareness of the applicability of the symbols to the realm. We say that our knowledge is true when we are clearly aware of this applicability. But this symbolic adequacy, as was intimated in chapter V, cannot always be determined with ease. Specifically the difficulty of its ascertainment may be due to the obscurity of the occurrents themselves, the indefiniteness of the symbols, or the complexity of the operational route which determines the applicability of the symbols to occurrents. Difficulties of the first kind are the result of two facts: (*a*) some occurrents are inherently obscure, e.g. high concretions and abstractions, highly extended and minute occurrents, the past and future, relations, etc., (*b*) other occurrents while not inherently obscure often require technical devices for their production in awareness, from which it follows that we cannot always be immediately and clearly aware of the entire realm of occurrents. Difficulties of the second kind are the result of the tendency to use symbols without a clear recognition of their operational derivations. Difficulties of the third kind arise out of the fact that we do attempt to talk intelligibly about obscure occurrents in spite of their inherent vagueness, and as a consequence do not always see the relevance of our symbols to their situations.

Accordingly, one must not expect to look about him and find an instance of perfect knowledge, if by this he means a system of symbols which adequately and completely represents the realm of occurrents or even a very limited portion of that realm. The obvious fact is, on the contrary, that existing systems are more or less adequate. Immediately the suggestion arises that it would be possible to evaluate existent systems upon the basis of the degree to which each of them meets the general demands of symbolic adequacy. Clearly such an arrangement would be impossible unless one were able to determine in advance some of the criteria for the measurement of this adequacy. Yet these criteria could be ascertained only by a consideration of all symbolic systems which lay claim to such adequacy. A serial arrangement might then enable us to see whether there are progressive difficulties in the attainment of knowledge, and hence how knowledge might advance by gradually overcoming these difficulties. This would suggest the possibility that knowledge might actually pass from a stage of less adequacy to a stage of greater adequacy. It would also suggest, perhaps, the *inevitability* of this passage to higher adequacy; i.e. there might be revealed certain features of knowledge which would enable us to say that an inadequate symbolic system is a relatively unstable system and tends by its own inner development to become more adequate. If this were possible, it should also be possible to see in each stage both remnants of the earlier and inferior stages and anticipations of the later and superior stages.

In essence, then, it is the task of this chapter, first, to determine what is meant by symbolic adequacy; and, second, to show how the various kinds of symbolic systems can be arranged serially on the basis of this conception.

Criteria of Adequacy

The general feature which is revealed with varying degrees of success by symbolic systems is the representative character of the symbolic complex. Such a system claims to be *about* the realm of occurrents. This claim, as we have seen, is not external to the system; the reference of the symbols to the occurrents is part of the meaning of the symbols. The adequacy of the symbolic system will be found in the accuracy with which the meaning contents of the symbols represent the actual contents of the occurrents. It must be emphasized here that the accuracy of a symbolic system does not demand that to the formal characters of the symbols there should correspond formal occurrents. We have seen that the form of a symbol is its way of referring to occurrents. But the adequacy of the symbolic system does not demand that there should be universal occurrents which are the referents of concepts, or that there should be facts, as entities different in kind from complex occurrents, which are the referents of propositions. Thus even the most adequate system of word symbols will lose something of the occurrents, viz. their particularity, and will add something, viz. the generality of symbolic reference.

But within these limits it is possible to state in comparatively simple terms what constitutes the adequacy of a symbolic system. Such a system will be adequate if it contains descriptive symbols for the representation of clearly given occurrents, correlational symbols for the representation of structures of occurrents (which are usually more obscurely given), and suppositional symbols for the representation of obscurely given occurrents. It follows that occurrents may be symbolized either directly through descriptive symbols or indirectly through suppositional symbols; in the former case the occurrents are clearly given, and in the latter they are not. It also follows that structures of occurrents may be symbolized through correlational symbols, which are then either descriptive or suppositional depending on whether the structures are clearly or obscurely given. Let us discuss each of these aspects of symbolic adequacy.

(1) An adequate symbolic system will exhibit a one-one correlation between clearly given occurrents or kinds of occurrents and descriptive symbols. (It is necessary to include " kinds " of occurrents as well as occurrents, since characterizing symbols, as differentiated from icons, refer to occurrents *generally* and not *specifically*). In an adequate system there will be a symbol for every occurrent or kind of occurrent, and for every occurrent or kind of occurrent there will be a symbol. Furthermore there will be not merely this formal isomorphism; it will also be true that the content of the symbol will reveal the content of the occurrent. Hence the significance of the symbol will lie not merely in its formal correspondence with the realm of occurrents but in the fact that the symbol is a substitute for the occurrent and enables us in the absence of the occurrent to refer to it determinately. The adequacy of each kind of symbol, e.g. names, icons, or characterizing symbols, is to be estimated in terms of its own specific manner of referring; hence it will

not be correct to measure the adequacy of names by their pictorial value nor of characterizing symbols by their capacity to point.

(2) An adequate symbolic system will exhibit, further, a one-one correlation between kinds of occurrential structure and correlational symbols. Here the reference of the correlational symbols is to kinds of occurrential structure rather than instances, since the specificity of the reference is determined by the terms thus related. In other words, correlational symbols never occur by themselves but always along with symbols for elements, the whole constituting a more complex symbol. This might have been inferred from the first criterion for adequacy. If complex occurrents (i.e. relational complexes) can be given clearly, and if there is a one-one correspondence between symbols and occurrents, then there must be complex symbols for complex occurrents as well as simple symbols for simple occurrents. An icon will mirror relations as well as elements. A concept which represents a complex will have as its equivalent a proposition asserting the complex, and the concept will reveal the complexity of its referent less adequately than will the proposition.

But a symbolic system will also be adequate structurally if it contains symbols for obscurely given structures as well as for clearly given. An obscurely given structure of occurrents is one in which one of the occurrents, or the relation (or both) is given obscurely. Symbols for obscurely given correlations become suppositional symbols. If an occurrent is given clearly but its structural relation to another occurrent is not given clearly, we employ a correlational symbol of a suppositional character to designate the relation. An adequate symbolic system will contain symbols for all such kinds of structural relations, and will unite these symbols with symbols for the clear or obscure occurrents which they connect.

(3) An adequate symbolic system will exhibit a one-one correlation between obscurely given occurrents or kinds of occurrents and suppositional symbols. However, although the suppositional symbols refer to the obscurely given occurrents, they will not have their meanings defined in terms of these occurrents. On the contrary, they will have their meanings

defined by operational routes from the clearly given occurrents through the medium of correlational symbols. By means of suppositional symbols we endeavor to grasp those things which are hinted at but not presented indubitably. But, if we are to talk about them, we must do so meaningfully; hence their reference is through structural relations to clearly given occurrents, and they are thus defined. Consequently an adequate symbolic system will contain symbols for all high concretions and high abstractions, all extensive and minute occurrents, all past and future occurrents, all remote spatial occurrents, all psychical, all supernatural, and all unreal occurrents. Every such symbol will be suppositional in character and can be used meaningfully only if it is defined through a specifiable operational route which is itself presumed to be based upon an actual occurrential structure.

Another way of describing this isomorphic adequacy of symbolic systems is in terms of the notions of unique and structural contents. Only descriptive symbols will have unambiguous unique contents, for these alone can be defined by a simple gesture of pointing. But all symbols will have structural contents. Then if descriptive symbols refer extensionally to occurrents and intensionally to other symbols which presumably also refer to occurrents, we may say that a symbolic system is adequate only if it is both extensionally and intensionally adequate. If it is extensionally adequate it will contain symbols possessing unique contents which refer to the clearly given occurrents; and if it is intensionally adequate, it will contain symbols for the structural relations of these occurrents to other occurrents, both those which are clearly given and those which are not. Hence an adequate symbolic system will point directly to the clearly given occurrents and indirectly to those which are obscurely given; but it will point to the obscurely given occurrents through intension, and thus we shall be able to see how these occurrents are related to the more obvious ones and therefore how we shall be obliged to set out in order to determine *where* they are and *what* they are.

The Development of Knowledge

But a system which exhibits all of these features of adequacy does not arise full grown; on the contrary, it is the product of a long period of development. Its temporal origin is hidden in obscurity, and its temporal development even in comparatively recent times is known only in vague outlines. Fortunately it is not our task to trace this evolution. There is reason to believe that it is in essence such as Comte described it to be.[1] It seems clear that knowledge must have had its origin in the things which were used, felt, and enjoyed. But this narrow realm must have yielded rapidly to expansion; as man's interests and capacities increased there must have been a progressive extension of the realm of the given to include more and more remote objects, more and more obscure objects. It was natural that man should first define these obscure objects in terms of himself, and it was perhaps equally natural that with his increasing knowledge of the uniformity of the world he should then define them in terms of abstract powers and capacities. It may even be that he suddenly realized the futility of all such modes of interpretation and decided that the ultimate task of knowledge must be not that of explaining but merely that of describing.

But if such has been the actual progress of knowledge, there is reason to believe that Comte was wrong in supposing that the positivistic attitude represents the ultimate stage in the development of knowledge. When a science has reached the positivistic stage, it has not reached maturity; on the contrary it is only beginning to be a science. Prior to this stage it has been superstition. The trouble with all knowledge prior to the positivistic stage was that it was not based upon a clear recognition of the possibilities of legitimate symbolism. It was thought possible to construct a science out of vague, indefinite, and obscure symbols. The positivistic stage is important, for it recognizes the futility of this aim. But positivism is wrong in supposing that we can remain satisfied by relegating all obscure entities to the rubbish heap and talking only about those things which are clearly and indubitably given.

[1] August Comte, *Positive Philosophy*, trans. Harriet Martineau (London, 1893), Vol. I, pp. 1-4.

The obscure occurrents demand recognition, just as they did for primitive man. But we have found a new technique for handling them. We simply refuse to say any more about them than we are entitled to say upon the basis of the way in which they exhibit themselves in the world. If they occur in certain situations, then it is part of their contents that they do occur in just such situations. And if these situations are seldom and irregular, this also is part of the contents of the obscure entities. But if the situations are repeated and uniform so that we become convinced of the inevitable connection of the obscure occurrents with these situations, we then have precisely the technique for removing their obscurity. For we know where they can be found, and we presume that they will be related to the situations after the pattern by which clear situations are related among themselves; hence we have operational routes for locating them and determining their character through the more obvious occurrents with which they are related. Thus the fact of their obscurity does not prevent their being talked about in scientific terms, i.e. through legitimate symbolism.

It seems reasonable to conclude, therefore, that with a recognizable technique for handling obscure occurrents we need not refuse to admit them into our scientific systems. But we must not admit them incautiously or without due examination. And we must recognize, even more than before, the need for an adequate system of symbols representing the clearly given occurrents. For we are to define the obscure occurrents in terms of them.

As a result, we may say that a symbolic system arises in a definite locus, develops according to a definite plan, and aims at a definite goal. The locus will be the range of clearly given occurrents or such a part of this realm as happens to enter first into awareness. The plan will be the progressive expansion of this realm to include all other clearly given occurrents, and still other more obscurely given occurrents which are related to them through spatial and temporal ties. The goal will be the adequate representation of this complete realm through a one-one correlation of descriptive symbols and clearly given occurrents, correlational symbols and occurrential structures, and suppositional symbols and obscurely given occurrents.

Such a formulation of the problem affords a very convenient principle of dichotomy according to which we may divide all symbolic systems into two groups. The principle is as follows : Is the system mainly concerned with the adequacy of its descriptive symbols, or is it mainly concerned with the portrayal of occurrential structures and obscure occurrents ? In an alternative formulation, is the system mainly concerned with the representation of those occurrents which are clearly and definitely given, or is it mainly concerned with the extension of that realm through the portrayal of those things which are merely hinted at ?

The application of this principle gives rise to a contrast which is usually expressed as the distinction between descriptive and explanatory knowledge, empirical and rational knowledge, real and ideal knowledge, inductive and deductive science, existential and non-existential science, *a posteriori* and *a priori* knowledge, etc. The sharpness with which the original principle is applied determines whether the distinction is to be considered one of degree or of kind. Usually it is considered to be one of kind.[1] I shall argue against this position. There seems to be an important distinction expressed by these terms, but the differences between these two kinds of knowledge are relatively unimportant as compared with their resemblances. The critical word in the application is the word " mainly." We shall find when we come to apply the criterion to actual systems of symbols that there are no sciences which are concerned wholly with the realm of the clearly given, and there are no sciences which are concerned wholly with structures and obscure occurrents. Thus there are no empirical sciences which contain no foretaste of rationality, and there are no rational sciences which contain no aftertaste of the empirical. Furthermore, since the plan for the development of knowledge calls for a progressive

[1] In C. J. Keyser's earlier and highly suggestive book (*Thinking About Thinking* (New York, 1926)) he makes the distinction essentially one of the degree of the development of knowledge, but in his later and less valuable book (*Pastures of Wonder* (New York, 1929)) he makes the distinction into one of kind. What is worse, he calls all deductive knowledge mathematics and all empirical knowledge science. This has been somewhat the fashion in recent years, but it seems to me to be a thoroughly vicious use of terms.

expansion of the realm of the given, there will be presumably a continuous temporal development in any one science as it passes from the clearly given into the obscurely given. We shall see that this development is inevitable—that a purely descriptive science represents an unstable science which will attain its stability by paying increased attention to structures and suppositional entities. A rational science may then be seen to be the result of the " postulational treatment of empirical truth."[1]

But within those sciences which may be called descriptive or empirical there arises another important distinction. This contrast is usually expressed as the distinction between pictorial and non-pictorial knowledge, representative and non-representative knowledge, image and imageless thought, and (erroneously) concrete and abstract symbolism. The contrast does not imply a distinction between true or adequate symbolism and false or inadequate symbolism, but rather a difference between forms of adequate symbolism. In both cases we are trying to portray as directly as possible the character of the realm of clearly given occurrents. On the one hand we have icons, maps, diagrams, duplicates, photographs, models, images ; on the other we have words or other characterizing symbols. Here the adequacy is determined by the manner of representation which is characteristic of the symbol, and each type is adequate in its own way. But on closer inspection the iconic symbolism reveals a certain advantage ; a clear-cut image or photograph of an object seems to have a much higher informative value than a word description of it. The image is more precise as to detail, more direct in its manner of referring, and pictorial of the *Gestalt* in a way in which the composite word symbol is not. Hence we tend to employ words only when a more directly representative symbol is not available.

Summarizing this discussion, we may say that an examination of the question of the adequacy of knowledge reveals a standard of measurement through the application of which we may arrange all symbolic systems into a series. All such systems will claim adequacy but with different emphases. This difference

[1] This is the title of Chapter IV of Carmichael's *Logic of Discovery*. I have found this much neglected work very helpful in the formulation of my ideas on the development of knowledge.

in emphasis determines a new law of the three stages, which will begin precisely where the Comtian law stopped. It will indicate that a descriptive science is not the culmination of the scientific enterprise but the laying of the foundations for operational expansion into the realm of obscurely given entities. It will indicate further that sciences tend, through their own internal development, to pass beyond the descriptive stage and to take on the characterisitcs of rational structure and deductive organization.[1]

Briefly (since the next three chapters are to be devoted to this thesis) the nature of the development can be formulated as follows : The first stage may be called, as above, the pictorial or iconic stage. This is the stage of *models*. The break between this stage and the succeeding stages is rather sharp. In the later stages images are replaced (or supplemented) by words and other characterizing symbols. Here the representative character of the symbol is increasingly lost, at least so far as its direct pictorial value is concerned. The second stage may be identified as the stage of *description*. This is distinguished from the stages which follow by emphasis on the clear occurrents as over against correlations and obscure occurrents. Finally there is the third stage, in which we retain characterizing symbols as opposed to icons, but in which we attempt to represent, for better or for worse, the structural features of existence and those relatively obscure occurrents which are often required in order to make this structure intelligible. This may be called the stage of *explanation*. It introduces an indirect manner of reference which permits representation in advance of intuitive awareness, though not, we hope, in violation of it.

We may now state as a preliminary to the discussion of the next three chapters the main principles according to which these stages may be differentiated from one another. First there will be the principles which distinguish the stage of icons from the

---

[1] This thesis of the temporal and logical continuity of descriptive and explanatory science has been suggested to me by a number of contemporary writers of whom Carmichael, Keyser, and Rueff are the most important. There are suggestions of the same thesis also in Mill, *System of Logic*, book II, chap. IV, sec. 6, 7 ; Russell, *Scientific Outlook*, (London, 1931), p. 64 ; Eaton *General Logic*, pp. 576–83.

stage of characterizing symbols ; then there will be the principles which, within characterizing systems, distinguish descriptive knowledge from explanatory knowledge.

Pictorial Knowledge

The features of pictorial knowledge are determined by its claim to the maximum of adequacy in the representation of the clearly given occurrents. This means two things.

(1) Pictorial knowledge is concerned mainly with the adequacy of the symbolic system in its representation of *particular occurrents*. Hence an image system is adequate when for each individual occurrent there is an individual image. Only by this method can we represent the unique features of existence. An iconic symbol refers only to the occurrent which it aims to portray, and thus lacks the feature of general reference which is always found in characterizing symbols. There are no generic images, as Berkeley showed. An image which is apparently generic is really only indefinite, and an indefinite icon is a bad one. A pictorial symbol seems to gain generality either by neglect of the specific features of the occurrent referred to or, retaining the specific features, by the recognition of the resulting lack of resemblance between the icon and the occurrents. In either case, however, it becomes an unsatisfactory icon, for in the former case it adequately represents only a part of the occurrent and in the latter case it inadequately represents the whole occurrent. Only a characterizing symbol can be adequate and general at once. Also it is impossible to represent pictorially a *class* of occurrents unless one has a symbol with the same multiplicity as that of the class. Thus to represent a_1 and a_2 pictorially demands a symbol which is plural to the extent of containing an element for a_1 and an element for a_2 ; otherwise it is not an adequate pictorial representation. If nature contains many occurrents which are very much like one another, then a symbolic system which is adequate to the portrayal of nature must also contain many symbols which are very much like one another. Hence the function of icons is the pictorial representation of the realm of occurrents through particular symbols for particular occurrents.

In such a system the aim is a perfect one-one correspondence, with the *same* multiplicity in the symbolic realm as is found in the occurrential realm.

(2) Pictorial knowledge is concerned mainly with the adequacy of the symbolic system in the *directness* of its reference to occurrents. It is most adequate in that the derivation of symbols is immediate and the verification of symbols is simple. The act of deriving a symbol is the operation of duplicative construction, and the act of verifying a symbol is simply the act of comparing two occurrents. This is the sort of comparison which is impossible according to certain dualistic interpretations of the cognitive situation, for if knowledge is representative we can compare one image only with another and never with an occurrent. But according to the point of view taken in chapter V the inspection of images is of the same kind as the inspection of occurrents, i.e. it is the intuitive awareness of an occurrent without regard for its metaphysical status. Consequently if one knows how icons mean and if he is given an occurrent, he may easily form an icon; and if he is given an icon and its referent, he may easily determine the adequacy of the reference. The directness of the reference lies in the substitutional character of the symbol; the symbol may usually be put in the place of the occurrent without significant loss.

Non-Pictorial Knowledge

On the other hand, non-pictorial knowledge is characterized by the weakening of its claim to perfect adequacy in favor of its claim to simplification. This means two things.

(1) Non-pictorial knowledge is concerned mainly with the adequacy of the symbolic system in its representation of *kinds* of occurrents. Hence a non-pictorial system is adequate when for each kind of occurrent there is an individual symbol. Characterizing symbols refer through their features of generality. In the operation of generalization through which they are derived there is a loss of the individual content of the referent. (This does not involve a loss in the *unique* content, for to the unique content of the occurrent there corresponds a unique content of

the symbol; but it does involve a loss in the specific spatio-temporal situation). Thus from the examination of a characterizing symbol one cannot tell *which* occurrent is being referred to, for it is not part of the meaning of any such symbol to tell where or when its referent is to be found. But with this loss in specificity of reference there is a gain in simplicity. For the plurality of the symbolic system need not be so great as that of the occurrential realm. Differences between individual occurrents which are not expressible in differences in kind will not be capable of symbolism in the system. We have lost the one-one correspondence, but the symbolic system has acquired a synoptic value. We are now able to talk about things in general without talking about particular things. The presumed gain is that, relatively speaking, things in particular are unimportant.

(2) Non-pictorial knowledge is concerned with the substitution of a *different kind of adequacy*. The symbol now *characterizes* its referent, and no longer resembles it in any significant sense. In this sense the reference of characterizing symbols is indirect, for to pass from the word as a group of letters to the occurrent as a referent one must know that to the group of letters a certain meaning has been attached by social usage. This is a common source of error and of disagreement. For if the problem is one of the verification of the applicability of a word symbol, say " red," to a situation, it may fail of solution because the individual either is not able to discern the fact of the red occurrent being in the situation, or does not know that in the English language the word " red " means the particular color which he discerns in the situation. In the case of icons this difficulty is avoided, for one can " see " the similarity between the symbol and the referent. But this loss in directness of reference is associated with an important gain, for by means of characterizing symbols we may refer significantly to occurrents which are obscurely rather than clearly given, i.e. we may talk about occurrents of which we cannot form satisfactory icons. Thus the way is opened for the expansion of our system into the realm of suppositional entities.

Descriptive Knowledge

Within non-pictorial symbolism descriptive knowledge will be distinguished from explanatory knowledge by its predominant interest in the clearly given, as over against the obscurely given, occurrents and structures. This will involve two points.

(1) Descriptive knowledge is concerned mainly with the derivation of symbols from clearly given occurrents, and hence is interested primarily in the *establishment of isolated symbols*. Knowledge in this stage is closely tied up with the practical. One of the most obvious features of existence is the fact that there are *things* which we are compelled to manipulate in response to the demands of our bodies for food, clothing and shelter. One of the most successful ways of handling these things is to name them, talk about them, and describe their modes of behavior. This necessitates the invention of symbols by which we may tag objects and thus act toward them co-operatively. Knowledge in the descriptive stage does not neglect entirely the fact of structure; we are also interested in how things are connected. But it is only the most obvious structures which we attend to, and we feel no need for the introduction of symbols for conjectural or hypothetical entities which will increase the integrative character of the body of symbols.

(2) Descriptive knowledge is characterized also by a general *disinterest in the establishment of symbolic systems*. There is too much concern about things to permit an interest in structure and organization. Relations are characteristically difficult to grasp and enter into awareness only when there has been a fairly distinct awareness of the elements to be related. Hence symbols in descriptive knowledge will be characterized by a more or less complete lack of organization. Contradictory symbols may coexist because they have not been brought into juxtaposition; relations of independence prevail, permitting any element of the " system " to be varied without change in any other element; and extension is predominant over intension, i.e. a symbol is defined by a gesture of pointing indicating the referent, rather than by showing the structural relation of the symbol in question to other symbols having a more direct extension. By and large, our complex of symbols adequately reveals the elements of the

world which are clearly given, but it very inadequately represents those features which are obscurely given, be they structural or non-structural.

Explanatory Knowledge

On the other hand explanatory knowledge will be mainly interested in making clear those occurrents and structures which are seemingly required to make the clearly given intelligible, yet are not themselves clearly given. This will mean two things.

(1) Explanatory knowledge is primarily concerned with the representation of obscure occurrents, especially those concerned with structure; hence it is essentially interested in the *establishment of symbolic systems.* The disorganization of the descriptive " system " is seen to lie in the presence of gaps; these gaps must be filled in by interpolating and extrapolating upon the basis of that which is clearly given, and thus defining as adequately as possible those obscure occurrents which are presumed to occupy these gaps. Taking the accumulated knowledge as a sample, what may we presume as to the rest of nature? What may we say as to the character of those occurrents and structures which are not clearly given, in order that those occurrents and structures which are clearly given may be rendered more intelligible? Hence explanatory knowledge exhibits a relatively high degree of integration, which in its ideal form constitutes a deductive system. There are no contradictions, few relations of independence, and a primary emphasis upon intension as against extension.

(2) Explanatory knowledge is characterized by a general *disinterest in the adequacy of isolated symbols.* Hence there seems to be a sense in which higher integration is attained only at the expense of descriptive value. This is true, but only in a limited sense. The symbols are defined through intension rather than through extension, but this does not mean that they are without extensional reference. On the contrary, it means that the obscurity of the occurrents referred to prevents their symbols being defined through extension; hence symbols are given meaning not by pointing but by showing structural relations to other symbols which can themselves be given meaning by

pointing. Consequently explanatory knowledge *seems* to be nonexistential, ideal, *a priori*. It is more properly characterized as existential, real, and *a posteriori*—but as referring to occurrents which are given obscurely and are thus incapable of being talked about until, in the exploration of the realm of the clearly given, we come upon occurrents with which they are related and in terms of which they are to be defined. The non-existential, the ideal, and the *a priori*, are only the existential, the real, and the *a posteriori* which have not yet been operationally defined.

CHAPTER XI

MODELS

In this chapter we shall attempt to determine the adequacy of pictorial knowledge. Regardless of the stage which this type of knowledge may occupy in the development of knowledge as a whole, it is clearly a kind of knowledge making a certain claim to adequacy. It is therefore relevant to examine its structure, describe its manner of referring to occurrents, and attempt to determine therefrom whether it has a better claim to the position of the most adequate form of knowledge than has characterizing knowledge. Thus it will be important to list its advantages and disadvantages, and to point out by comparison the correlative advantages and disadvantages of non-pictorial knowledge.

The relevance of this problem to the general subject of the logical structure of science is great, and is acknowledged by many authorities.[1] The problem is usually stated in terms of the contrast between the pictorial and the conceptual formulations. Do we think in images, models, pictures, or do we think in mathematical symbols, abstract conceptual schemes, word-representations? As Aliotta points out,[2] investigators are rather sharply divided into two schools on the issue. On the one hand are those who insist upon the necessity for and adequacy of models; in this group are to be found most of the English physicists: Faraday, Kelvin, Lodge, Maxwell. On the other hand are those who argue for the inadequacy of images and insist upon the necessity for employing abstract representations; in this group are to be found Mach, Ostwald, Duhem, Hobson, Poincaré, and others. Duhem[3] insists that the difference is based fundamentally upon the type of ideation employed by

[1] Hobson, *Domain of Natural Science*, p. 82; Bridgman, *Logic of Modern Physics*, pp. 52–60; Enriques, *Problems of Science*, pp. 363–6; A. Aliotta, *Idealistic Reaction Against Science* (London, 1914), part II, chap. V; Cassirer, *Substance and Function*, p. 141; Cohen, *Reason and Nature*, pp. 72–75; Eddington, *Nature of the Physical World*, pp. 209–10; P. Duhem, *La théorie physique* (Paris, 1914), chap. IV; Feigl, *Theorie und Erfahrung in der Physik*, pp. 94–98.
[2] *Op. cit.*, p. 390.
[3] *Op. cit.*, pp. 77–81.

the investigator, and Poincaré[1] suggests that the differences may be nationalistic. But in any case the issue is important, for it has to do with the kind of meaning which is most fruitful in scientific symbolism. There is clearly a difference in the way in which images and characterizing symbols refer to occurrents, and the important question is to determine which has a greater claim to adequacy, both as a means of conveying information about what is not clearly understood, and as a means for furthering scientific discovery. It is to this problem that we shall address ourselves.

It must be understood that we are speaking of imagery in the broadest possible sense.[2] As a consequence actual material models, constructed out of string, rubber, pulleys, wire, etc. become a special case within the general treatment. The construction of materal models for thought images is simply an attempt to overcome the fluctuating character of the symbol and does not change the problem in its essential details. In both cases the symbol is concrete, i.e. an occurrent. Hence the problem of meaning is that of determining the degree of resemblance between two occurrents, one of which functions as a symbol and the other of which functions as that which is being symbolized. The popular presentations of recent science which have as their aim the making clear of the obscurities of science through the construction of working models, e.g. of the astronomical system, of the structure of the atom, of the action of electrical charges, and the like, all perform the same function for the uninitiated individual as does the concrete imagery of the investigator himself for his own understanding. Elaborate instrumental techniques have enormously increased the possibilities of pictorial symbolism. Our problem will be formulated without regard to the specific kind of icon which is being employed. Pictures, diagrams, sketches, maps, outlines, etc. will belong in the same category as elaborate working mechanical models. Their essential identity lies in the oneness of reference;

[1] *Foundations of Science*, pp. 3–7.
[2] Not, however, so as to include word symbols. Duhem (*op. cit.*, pp. 110–12) and Peirce (*Collected Works*, Vol. II, par. 279.) consider algebraic signs as images. I should not.

they all attempt to symbolize through some form of mirroring, more adequate in some cases, less in others.

The problem of the rival claims to adequacy of these alternative methods is much complicated by the fact that there is probably no situation from which both pictorial and non-pictorial forms of symbolism are absent. Man does not think *either* pictorially *or* non-pictorially; he usually thinks by both means. Psychologists have debated the question as to whether imageless though is possible.[1] The issue is not important for our purposes. There seems to be some reason to believe that thinking in words, i.e. word-meanings, is impossible entirely apart from images of some kind, though the images may be merely those of the *words* rather than those of the objects referred to. Furthermore, the question as to whether there is ever " thoughtless imaging " is simply a question of whether intuition, as defined in chapter V, ever occurs apart from the awareness of word symbols. I suggested at that place that the issue was not important; what is important is that these are two different kinds of awareness, which can be distinguished from one another, even though they always occur in association with one another. Such seems to be the significance of the problem here. If, as seems likely, imagery always occurs along with word symbols, and if, as seems equally likely, some word-meanings are found in the presence of all imagery, it is then vital to determine in any situation which is the *predominant* symbol, i.e. which symbol we are employing for the purpose of obtaining information about the occurrent. For example, in thinking about molecules one may find present in his awareness, on the one hand, a concrete image of small billiard balls in a bushel basket, bounding against one another, and, on the other hand, the abstract meaning of the equation

$$pv = 1/3 \, nmu^2$$

The problem of scientific methodology is, then, one of determining whether the scientist makes predictions as to the further character of molecules by a projection of the features of the image, or by a deduction of the consequences of the law. Usually he cannot

[1] W. James, *Principles of Psychology* (New York, 1901), Vol. 1, p. 243; G. F. Stout, *Manual of Psychology* (London, 1921), pp. 251–3.

do both at once, for the mathematical features may be such as could not be readily pictured, and the pictorial features may be such as would not permit the direct derivation of a complicated mathematical characterization. He may, and usually does, employ both methods, but alternately, so that the notion of molecule is given increased content through supplementation by images and by abstract symbolism. Often, as Duhem points out,[1] discoveries apparently arise from images, but only because the investigator forgets the abstraction which alone enabled him to make the model in the first place.

If we stop to examine the claims of iconic symbolism, we shall see that there is something inconsistent about its claim—a fact which may account for its general inadequacy as compared with characterizing symbolism. The manner of meaning is resemblance. Hence the adequacy of an icon is determined by the degree of success with which the image reveals the content of its referent—as to elements and as to structural organization. Perfect duplication, therefore, is the ideal and aim of every iconic system. But if this ideal has been attained, what about the adequacy of the symbolic system from another point of view? A symbolic system seems to have at least two functions: (*a*) It gives us a substitute for that which is referred to so that we need not intuit the referent everytime we wish to learn something about it. (*b*) It simplifies the referent and thus gives us something more convenient than the referent when we are concerned not with it as a whole but with selected features of it. The first of these demands seems to be met by an image system, and we shall consider it immediately. But the latter is not met. In fact, just to the extent to which the image system becomes adequate on its representative side, it becomes inadequate on its simplifying side. For a perfectly adequate image would be as complex as the referent itself, and there would therefore be no reason for employing the symbol instead of the occurrent. Thus every image system contains a principle of dissension; it endeavors to portray accurately, yet it endeavors to portray by simplification, i.e. by omitting.

[1] *Op. cit.*, p. 138.

MODELS 253

The reconciliation of these two points of view in the construction of a pictorial system is sought in two ways, neither of which is satisfactory. On the one hand, there is the deliberate attempt to make the image *general* by making it *vague*. This is akin to the attempts to form composite images by so superposing images upon one another that only the outstanding features remain relatively distinct. Now, although such an image does represent a simplification of its referents it does so at the expense of its main claim to adequacy. For just to the extent to which it becomes vague it becomes a bad symbol; it must represent precisely or it doesn't represent at all. That which is referred to is definite; hence any representation of it must also be definite. So also with the replacement of models by pictures, and of pictures by designs, and of designs by sketches. With each substitution made for the purpose of simplification there is a loss of detail which renders the symbol inadequate. An icon cannot be a simplification of its referent and at the same time refer to it successfully in the manner characteristic of this type of symbol. On the other hand, there is the attempt to make the image definite and specific as to detail, and thus precise as to content, but at the same time to see the definiteness of detail as not part of the *meaningful* content of the symbol. For example, one can use a particular triangle as an image of triangles in general by seeing that the shape and size of the specific triangle are not part of its meaning; this involves giving a specific image a general reference and making the general reference not dependent upon the specific character of the symbol. This can be done, and probably often is done in actual symbolic situations. But if it is done, then one is forsaking iconic symbolism. For if a triangle becomes a symbol for triangles in general, and if the content of the meaning reference of the symbol is not revealed in the character of the symbol, then the symbol is not an icon at all; it becomes a characterizing symbol. Thus one could draw a triangle whenever he wished to refer to a triangle, and the symbol would then refer to triangles in the same way as does the word " triangle," except that, as in the case of onomatopoetic words there would be a rough similarity between the occurrent and the physical symbol. This would not make it into an icon, however, for the significant reference of the symbol would be determined, not by its pictorial representation, but by its social uses.

The result is that the adequacy of a pictorial system must be measured in terms of the degree to which it can retain its accuracy in the portrayal of detail while at the same time extruding some elements of detail in the interest of simplicity. There are many ways in which this can be done. One of the most important ways is in the matter of size. Clearly if the occurrential situation is of very wide or of very narrow extent, we can gain in our representation of it by changing its dimensions. We construct small models of large systems, and we construct enlarged models of small systems. We do this, realizing the inadequacy of our image in the matter of size, but hoping that this inadequacy will not destroy the general pertinence of the representation. So also we represent long durations by speeding up, in our model, processes which occur very slowly in nature; and, similarly, we represent short durations by slowing down in our model, processes which occur very rapidly in nature. (The motion picture machine has proved to be an important instrument of iconic symbolism in this regard). Further simplification may be obtained by neglecting features associated with the occurrent but of no concern to us in our attempt to represent any other of its aspects. For instance, if we are constructing a model of the astronomical system we may neglect the differences in size among the heavenly bodies, since we are interested merely in showing how they move. Similarly we may neglect the stuff of our system and represent it in terms of an entirely different stuff. Or we may neglect entirely the fact of a third dimension and represent the occurrent through a projection upon a plane. So, detail may be lost when we neglect differences in color and employ a draftsman's design. But when we have reached symbols of this kind, there is a more or less complete loss of adequacy; the symbol ceases to perform its symbolic function; it represents either by becoming vague or by demanding that its elements be seen not in their direct pictorial function. In either case it ceases to be an icon.

Of course, it should be recognized that even a system of word symbols possesses a certain iconic character. We have insisted that the adequacy of a system of characterizing symbols resides in its capacity to represent not only the elements of the world of occurrents but also its associative and dissociative relations; furthermore the structural relations between the symbols must correspond to the structural relations between occurrents. There is a structural and contentual similarity between the world of occurrents and the symbolic system. This makes the word system pictorial of the realm of occurrents. But the nature of the reference in the two cases is different, in spite of this important similarity. In the case of icons a knowledge of the content of the symbol, plus a knowledge of the general form of iconic reference, enables one to locate the referent. But in the case of characterizing symbols, the referent can be determined only by a knowledge (1) of the symbol (i.e. the word as a physical entity), (2) of the general form of characterizing reference (i.e. conceptual, propositional, correlational), and (3) of the specific form of the reference (i.e. of the content of the *meaning* of the symbol). Each word symbol, as we have seen,[1] has a specific way of referring to occurrents; this is a diversification of the general form of reference which is common to all symbols of the type, and varies from individual symbol to individual symbol. The specific form of reference is tied up closely with the specific character of the word as a physical entity, but it is not directly revealed in this character as is the case with icons. Hence we are not able to say that words mean occurrents in the same direct way that images do. Furthermore the variation in the kind of reference of characterizing symbols destroys the directness in the picturing. Correlational symbols look like concepts, and concepts (at least complex ones) look like propositions, but the manner of referring in each case is different. Thus it is only through the general and specific meanings of word symbols that their referents can be determined, but it is through direct inspection of icons that their referents can be determined. This may perhaps be expressed crudely by saying that the referents of icons can be determined by an inspection of their physical form, while the referents of words can be determined only by an inspection of their logical form.

[1] Chapter VII, pp. 141–144.

256 THE LOGICAL STRUCTURE OF SCIENCE

Advantages of Pictorial Symbolism
It is time, perhaps, to consider the advantages, if there be any, which accrue to this type of iconic symbolism. In the preceding chapter we discussed two of these advantages. There we saw that the pictorial system enables us through symbols to portray in as great detail as the complexity of the system permits the features of the *particular occurrent*. Thus icons enable us to grasp individuality in a way in which characterizing symbols do not. Furthermore, we saw that pictorial symbols have the advantage of direct verifiability. Through a mere examination of them, knowing that they are pictorial, one gains information about their referents. This is the great advantage of models for the conveyance of information; one need only examine the model. He is not obliged first to learn the meaning of words, and then from the system of words to pass to the referent.

We referred above to the substitutional character of a pictorial system. This is a feature of all symbolic systems except those which are made up wholly of proper names or indices. We make symbols in order that we shall not be obliged to reproduce the occurrents in direct awareness on each occasion when we wish to examine them. Words also have this function. But images have a great advantage over words in this respect. Since images are accurate representatives of their referents, they act as adequate substitutes for them. For instance, if I wish to determine some detail of the façade of Notre Dame cathedral I may obtain it more quickly and more easily by referring either to my image of the cathedral or to some photograph of it than by turning to a book written about it. For when I get the word symbol, I still do not have what I want; what I desire is the detail of the referent, and this can be obtained only through the *meaning* of the word. A diary of one's travels in a foreign land is helpful in recalling the details of his experiences, but a vivid imagery is still more important. Hence, in the ease with which the symbol and the referent may be replaced by one another, pictorial symbolism has a distinct advantage over word symbolism.

Still another feature of imagery, which constitutes one of its advantages over word symbolism, is its flexibility. Not only can we make images more or less at will, we can also modify them freely in one or more of their features and then detect what changes seem to follow as a result in their other features. We can perform imaginative experiments by introducing transformations into images and then noting what further contents the images may take on in view of this modification. This procedure is in many ways analagous to physical experimentation; images yield to manipulation just as physical objects do. There is the same general kind of resistance to modification, and the same general kind of fixity of outcome under the influence of modification. Probably in the case of images both the degree of resistance and the degree of fixity in the outcome are less. We can modify images more easily and through a wider range of possible transformations, and we discover less constancy in the character of the outcome. The former is an advantage and the latter is a disadvantage. We may perform an imaginative experiment when conditions prevent us from engaging in a physical experiment. But when we have performed our experiment we find that the result is somewhat less satisfactory from the point of view of definiteness than if we had carried out the physical experiment. If images were perfectly pliable, this would constitute a serious disadvantage. But in spite of a certain flexibility images possess also a relative fixity. Hence, although we find that we may make images more or less at will, we discover that when we have made them they are of one kind rather than another. As a consequence they resist modifications except in so far as these entail a rather limited range of further modifications. Through this interdependence of transformations images acquire an important explanatory value. If we wish to know in advance of observation how a situation might behave under conditions which are a variation from the normal, we may form a concrete image of the situation, introduce the transformations, and note what further modifications occur in the image. It is clear that this procedure is not highly informative, and I presume no one ever supposed that by means of it one could foretell accurately the march of events. But it gives a suggested hypothesis with which to set up an experimental situation whose aim will be the proof or disproof of the result of the imaginative experiment. Thus manipulation of images is a fruitful source of working hypotheses. When these images are material models, this method increases in value. For if the model is a more or less accurate reproduction of the original, the result of any modification introduced into the model ought to correspond to the result of a similar modification introduced into the original. Hence we may say that images are valuable instruments for the enlargement of our knowledge of a given situation, not in the sense that by a mere inspection of the image we may learn about the more inclusive situation, but in the sense that through the symbol we may experimentally extend the situation in time or space, or introduce modifying factors, and then, noting what happens in the imagery, return to the situation with a point of view for further observation. Thus images are important aids to the hypothetical method.

Disadvantages of Pictorial Symbolism

When we turn to a consideration of pictorial knowledge from the point of view of its inadequacies rather than its adequacies, we find that there is more to be said.

Two of these inadequacies have already been treated. In the previous chapter we suggested that icons tend to be replaced by word symbols because of the incapacity of the former to portray *kinds* of occurrents, or to refer to occurrents generally rather than individually. Pictorial systems lack the simplifying function of word systems, for there is no way of portraying spatio-temporal diversity within qualitative similarity except through the employment of a complex symbol exhibiting the same spatio-temporal diversity as the system portrayed. But by means of a single word we can represent a spatio-temporal distribution of qualitatively similar occurrents.

Furthermore we saw that the indirect reference of word symbols, although it involves a loss in the immediacy of verification, gains a certain advantage in the portrayal of occurrents designated by suppositional symbols: it enables us to talk about occurrents in advance of any clear awareness of them. Now the important caution to be kept in mind in this advance awareness of occurrents, as we saw in our discussion of the hypothetical method, is to attribute to the suppositional symbols only such content as they can either be inferred to possess by the constructional method or supposed to possess through the hypothetical method; i.e. we must not make them more determinate than we are entitled to on the basis of the data. Here the inadequacy of pictorial symbolism can be readily seen. Images must be determinate if they are to be good images. Hence, if our suppositional symbol is an icon instead of a characterizing symbol, we are almost certain to attribute to it more content than we are entitled to. This is the danger in thinking of molecules as billiard balls, forces as rubber tubes, ether as a fluid, causal action as compulsion, gravitation as a pull, disease as something which is carried, growth as an urge, etc. Our concrete imagery may lead us astray for we unconsciously read into the occurrent all of the specific features of the image.[1] Having made the symbol highly determinate as to content, we demand that the exemplifying occurrent possess the same determination. The danger in this lies in the fact that, should we fail to find an occurrent exhibiting this high determination, we reject the suppositional symbol as being inadequate; whereas what we should do is to reject only that part which has been added unconsciously as a result of the over-concrete image. If we keep the suppositional symbol relatively indeterminate there is a greater probability of finding the occurrent. Then, having found it, we may fill in its details empirically. Hence, we are at a disadvantage when we try to symbolize hypothetical entities in terms of pictures; images must be taken or left as concrete wholes and cannot be built up by a step-by-step process. But, in the use of word symbols, by the progressive addition of word characterizations we may fill in the content of the suppositional symbol gradually, and with a recognized inferential ground in each case if such may be found. Accordingly, we do not demand more than we can legitimately expect experience to reveal to us.

The most obvious limitation of image systems is their incapacity to portray in any adequate sense the obscurely given occurrents, i.e. those occurrents of whose existence and character

[1] Cf. Russell, *The Scientific Outlook*, (London 1931) pp. 69, 85; Eddington, *Nature of the Physical World*, p. 275.

we cannot be highly certain, because they cannot be clearly and distinctly intuited. Included in this group will be all high abstractions—such as occurrent, quality, relation, time and space—and all high concretions where the associative complex is too great to be adequately pictured in all of its details. Images will also be inadequate as representations of the past and future, though in this regard they are, relatively speaking, very adequate. At least they will be adequate if the events which we are attempting to portray are not such as would be given obscurely if they were given in the present (e.g. high concretions and high abstractions). The success of icons in the portrayal of the past and the future lies in the fact that the image may represent the occurrent exactly *except for the time feature ;* this time feature cannot be represented for the image is always in the present but that which is referred to is in the past or the future. If one insists that this is an important feature of the past and the future, then he must admit that imagery is inadequate. Included also in the group of obscure occurrents will be the highly extended in space and time, and the highly minute, e.g. extents smaller than about $1/32$ of an inch, and durations shorter than about $1/10$ of a second. In all of these cases we attempt to symbolize in a precise image something which is inherently vague.

One may insist that, if there are such occurrents as these, they can be imagined. For, if they could not be perceived they could not be known to exist at all ; and if they can be perceived they can be imagined. But the point is not whether they can or cannot be perceived ; it is simply a question of the clarity with which they can be perceived. Certainly none of these occurrents can be perceived as clearly as can, say, a small red ball ; hence none of them can be as clearly pictured in imagination. We *try* to picture them in imagination. For example, we try to imagine the minute by constructing an image on a larger scale, but this destroys the representative character. Or we try to picture pure length by imagining a very thin line, but this again is a bad image. Or we try to think of a high concretion by thinking successively of its elements ; but this is an inadequate image, for the occurrent is a coexistent whole not a succession of elements.

Thus if the function of our symbolic system is the distinct portrayal of what certainly exists, then an image system has a high measure of adequacy; but if the function of such a system is the portrayal of occurrents which are indefinite and vague as to content, and which may exist but are not certainly known to do so, then an image system is not the most adequate type of symbolic instrument. An image must be definite and it must refer definitely. But a word symbol may be used significantly, even though it has only a minimum of definite content. Hence word symbols are particularly well adapted to the representation of occurrents which are only obscurely known.

From this it follows that the forces which cause an iconic system to be superseded by an alternative system are inherent in the former. Man is obscurely aware of many things of which he wishes to be clearly aware. If he cannot be clearly aware of them in terms of the only available symbolism, he devises a new form. The urge to extend the symbolic system into the realm of entities which are merely suggested rather than definitely present is probably practical in its ultimate origin. So long as imagery is adequate to man's adjustment to his environment, there is little reason why it should be augmented or superseded. But, with the developing complexities of the organism and the parallel revelation of previously unknown features of the world, man finds himself more intimately connected with nature. If he is to survive he must know more about nature. Hence, he endeavors to reach beyond the immediately given and to adjust himself to that which is not immediately given but apparently efficacious in the control of the given. In this endeavor he abandons images and adopts word symbols.

Another disadvantage of images is their relative instability as compared with word symbols. This can be avoided in certain cases by embodying the image system in a material model which then possesses the same degree of permanence as that which is being represented. But in strict image systems, i.e. systems of "mental" images, which constitute the more important type of pictorial symbol, instability cannot be avoided. Images change in content even while we are examining them. Consequently a significant question to be raised when one is attempting to represent an occurrent by an image is : Which image ? A word of fairly constant meaning may suggest a variety of images. Word meanings undergo the same fluctuation, but not to so great a degree. Hence, in using words instead of images we symbolize with greater permanence.

It may seem strange that we are here emphasizing the instability of images as one of their disadvantages when we suggested a few pages back that the flexibility of images gives them an advantage over word symbols. But it can easily be seen that these notions are not incompatible. By flexibility in the former context was meant the variability in the content of images through the influence of definite acts of modification. Images are flexible because we can insert or extrude factors more or less at will and then examine the resulting symbols in view of this transformation. This gives images an advantage for by means of them we can become aware (at least through symbols) of hypothetical situations. By the fluidity of symbols in the present context, however, I mean the fact that an image which we presume to be constant is undergoing transformations in spite of our attempts to keep it fixed. This gives images a disadvantage for we have a multiplicity of distinct symbols all representing the same occurrent. Furthermore, this fluidity may be so great as to weaken to an extent the predictive claim of images. If images do enable us to predict—since they permit us to operate upon them experimentally and thus note what will take place in advance of the actual happening—they do so only to the degree to which they retain a certain rigidity under the influence of these transformations—a rigidity which can be attained only if we can agree upon certain rules according to which symbols ought to react to these voluntary modifications. In other words, if there is no constant way in which a symbol usually changes under the influence of the intrusion or extrusion of a certain feature, then the symbol is so fluctuating in character as to be a thoroughly bad symbol, and consequently it has no predictive value. Clearly, words have a superiority in this regard. There are no rules for the transformation of images, i.e. images have no syntax; but words do have a syntax so that we know in advance roughly at least, how a word system must undergo modification under the influence of the intrusion or extrusion of a symbol. And this syntax has a relative permanence. In short if images fluctuate to such an extent that there is no constancy in the way in which elements may unite to form complex symbols, then the flexibility of the symbol (its capacity to be subjected to imaginative experiments) is no longer any great advantage.

Duhem[1] refers to another aspect of image reference which constitutes an important disadvantage of this type of representation. The great variety of possible image-representations results in a diversity in the kinds of images which are employed. This is no disadvantage if the situations which are being portrayed are essentially different. But if a large number of different images is employed in order to emphasize various aspects of a complex entity, and if these images are incompatible with one another, then, although we see in clearer light the workings of the part, we are forever prevented from seeing the whole in operation. Duhem gives a significant list of the various images which Lord Kelvin employs to represent the ether[2] and then suggests that in order to get any definite idea of the constitution of matter one should somehow have to synthesize all of these images and then join to the result all of the images which Lord Kelvin does not himself employ but recommends to the reader as having been advantageously used by other physicists. The characters of these images make such a synthesis impossible. This is to be expected in view of the fact that every image takes on a greater concreteness than is demanded by the situation which it explains. It is the addition of these further features of concreteness that prevents the synthesizing of the image with alternative images of the same situation. Thus one can imagine a highly concrete situation only by a highly concrete image; if one tries to construct an image synthetically from the images of its aspects, he runs the risk of finding himself unable to fit the partial images into the total picture.

Another disadvantage of pictorial symbolism is closely related to the foregoing. Word symbols are more adequate than icons in the representation of fields where our knowledge is

[1] *Op. cit.*, pp. 118–20.
[2] *Ibid.*, pp. 121–2.

internally contradictory. One might well insist that the difficulties which are inherent in Lord Kelvin's portrayal of the ether are not features of the symbolism but the expression of a fundamental irrationality in the ether itself. In order to account for the great velocity of light we are obliged to attribute to the ether certain incompatible properties, viz., a minimum density and a maximum elasticity ; hence it must be at the same time a perfect fluid and a perfect solid.[1] More recently a similar problem has arisen in the reconciliation of the corpuscular and wave properties of light. Now it is clear that image symbolism does not help at all in cases of this kind, for we cannot form icons which are internally incompatible.[2] And it follows that if imagery is the essential mode of scientific symbolism the scientist is compelled to call a halt to all investigation when such a contradiction appears. If word or characterizing symbolism is employed, or, specifically, if mathematical symbols are employed, the difficulty recedes into the background or is synthesized into a higher abstraction which reconciles the incompatible elements.[3] But even if the difficulty is not resolved the important feature of word systems—at least of those which have not reached a high degree of integration— is that they can contain temporary incompatibilities without suffering complete disruption.[4] But images can never be internally contradictory at any level ; if thinking is carried on in terms of imagery, we simply cannot *think* things which are internally inharmonious. On the other hand, if thinking is measured in terms of word-comprehension we *can* think in contradictions. For we can assert in a proposition that two terms are contradictory. We can *say* that the attribution of corpuscular and continuous properites to light is contradictory. Such a symbol is not meaningless. In fact we have seen that there are kinds of propositions which assert relations of incompatibility between concepts ; these statements are meaningful and refer to repeated

[1] J. Tyndall, *Fragments of Science* (New York, 1898), Vol. II, p. 106 ; Pearson, *Grammar of Science*, pp. 289–92.
[2] Stace, *Theory of Knowledge and Existence*, p. 382.
[3] Bavink, *The Natural Sciences*, p. 191 ; Jeans, *New Background of Science* (New York, 1933), pp. 162, 189.
[4] Cohen in an article entitled " The Logic of Fiction" (*Journal of Philosophy* Vol. XX, p. 477) denies that fictions are contradictory. Cf. also Stace, *op. cit.*, pp. 439–42.

dissociations of occurrents. But they do not refer by picturing, for one cannot picture an incompatibility. To be sure, word symbols which state contradictions are not desirable elements of a symbolic system, and the gradual perfecting of the symbolic system necessitates their abandonment or their absorption into higher unities. But with word systems this can be done gradually and without employing symbols whose very character makes them unthinkable. Thus word systems give us instruments for carrying along contradictions which have not yet been resolved, but which, we hope, are to be resolved without a complete abandonment of the systems in which they occur.

Finally, image systems prove to be inadequate in the representation of the structural features of existence. Relations cannot be readily pictured. This is expecially true of those general kinds of relation which reveal the structural organization of occurrents, e.g. associations and dissociations, temporal orders, spatial and temporal inclusions. If the structure is considered in one of its instances, then there is not such a marked inadequacy, for the occurrents may in general be pictured and even the relational feature may be portrayed in a relational symbol. But we have seen that the important structures are those which are exhibited in *repeated* associations and dissociations. Certainly we cannot form an adequate image of a universally repeated association. Hence if we rely solely upon image symbols we shall not be able to distinguish between compatibility and implication, or between one-way implication and co-implication. But these are distinct kinds of structure, and to confuse them can only produce difficulty. It is probably impossible also to represent causal correlations adequately through imagery. The attempts to do so have resulted in anthropomorphism, i.e. the reading into the causal situation of such features of the image as the "power" of the cause and the "necessary occurrence" of the effect. It follows from these considerations that, as one attempts to develop his knowledge of occurrents in the direction of structural organization, he finds himself obliged to abandon more or less completely all forms of concrete imagery. If he chooses to retain iconic symbolism he must recognize clearly its consequent dangers.

The outcome of this chapter is essentially as follows : Icons are an important form of symbolism. But their adequacy is of a certain kind and their application therefore of limited range. If we are interested in simple reproduction of individual occurrents which are capable of being given clearly, we cannot do better than to employ pictorial symbolism. It contains a vividness, conciseness, and accuracy which cannot be improved upon. But as we recognize the importance of extending our symbolism to include kinds of things, vaguely and indefinitely given things, structures of things, we are compelled to admit the essential inadequacy of this form of symbolism. We therefore tend to replace it (or to supplement it) by systems of characterizing symbols which seem to possess precisely the features which we demanded of pictorial symbols but found lacking in them. Because this increased advantage involves also a loss in direct representational and substitutional value, we seem to be getting further away from actual occurrents. However, this separation is more apparent than real, for the nature of the separation is describable and actually constitutes a part of the meaning of the symbol. Thus, in a sense, the symbol means directly, though its route is through intension. But this involves considerations to be treated in greater detail in the following chapters.

CHAPTER XII

DESCRIPTION

In this chapter we shall endeavor to make a more detailed analysis of descriptive knowledge. The last chapter enabled us to see by implication certain of its features. It is a kind of non-pictorial knowledge and therefore has all of the advantages and disadvantages of this type of symbolism. Specifically it must forsake occurrents for kinds of occurrents and direct resemblance for characterization. There is thus a loss in adequacy. But there are important compensating gains in simplicity, in fixity, and in the adequacy of the portrayal of relations, contradictions, and obscurely given occurrents. Hence by and large we seem to be moving toward a more adequate system when we pass from iconic symbols to characterizing symbols.

All of this will be presumed in the present chapter. Here our problem is one of determining, within the scope of characterizing symbolism, the relative adequacies of descriptive and explanatory knowledge. Hence we shall be concerned more with the features of descriptive knowledge which distinguish it from explanatory knowledge than with those which differentiate it from pictorial symbolism in general.

The best approach is to refer back to the general criteria of symbolic adequacy.[1] A symbolic system is adequate if there is a one-one correlation between symbols and occurrents, or, as we expressed it, between clearly given occurrents and descriptive symbols, structural relations and correlational symbols, and obscurely given occurrents and suppositional symbols. This makes our symbolic system satisfactory as to clearly given things and relations, and as to those things and relations which seem to demand symbolism without offering themselves with sufficient precision to permit of direct portrayal. It is this distinction between the clearly given and the obscurely given that determines the features of descriptive knowledge. Fundamentally, descriptive knowledge insists upon the importance of constructing a symbolic system which is adequate in the

[1] Chapter X, p. 234 ff.

representation of clearly given occurrents and of those structural relations which are also clearly given. Secondarily, descriptive knowledge either denies altogether the legitimacy of trying to symbolize obscure occurrents or else relegates them to a position of logical inferiority, which is equivalent to saying that it is willing to attempt to symbolize them only if this proves to be a fruitful method for increasing the range of the clearly given occurrents. This determines the task of the present chapter. We shall try to estimate the adequacy of a descriptive system from three points of view: (1) Its adequacy as to descriptive symbols, (2) its adequacy as to correlational symbols, and (3) its adequacy as to suppositional symbols.

Before plunging into this discussion it will be well to anticipate a difficulty. Descriptive knowledge is relatively unstable and tends to pass over into explanatory knowledge. This factor of transition is inherent in the character of the descriptive system so that by its own development it tends to become something other than itself. In view of this fact we shall have great difficulty finding a case of pure descriptive knowledge, i.e., a stage in which the disruptive forces are quiescent rather than active. In general we shall find that the actual sciences, even those which are acknowledged to be empirical in character, exhibit many features which entitle us to designate them rational sciences. Hence we shall have to extrapolate from such sciences in order to estimate what would be the character of a purely empirical science. The risk which we shall run on this score will not be great, since our task is that of showing the essential relativity of the distinction between description and explanation. Thus, if we can find clear-cut cases of sciences which are more empirical and sciences which are less empirical, and if we can show that the distinction is one of primacy of interest rather than of kind of subject-matter or kind of method, we shall have made our point clear.

Adequacy of Descriptive Symbols

A descriptive system is, in the first place, adequate as to its symbols for clearly given occurrents. This means that there are symbols for all such occurrents and they are derived from these occurrents by the most direct sort of operational route. Clearly there are two predominant interests characteristic of a science in this stage; these are in the collection of data and in the symbolizing of objects collected. Consequently the impelling spirit of this stage is that of discovery and the accumulation of data. Knowledge can progress only by the rapid acquisition of information. Description represents science in the exploratory stage in which data are being gathered, and in the classificatory stage in which they are being grouped and named. It is the stage in which man as a pure empiricist *passively* " listens to nature," but as an incipient rationalist *actively* begins to ask nature questions, to prod her and to force her to yield up her secrets. The result is often an accumulation of facts which is much in advance of their interpretation and organization. But at any rate the facts are there, and it is the essential function of descriptive science to reveal the facts. Hence in order to discuss this phase of science, we must consider those operations which make possible the discovery of occurrents and the formation of descriptive symbols. These will be of three kinds : (*a*) physical or manipulative operations, (*b*) selective operations, and (*c*) inventive operations resulting in the actual formation of the symbols.

(*a*) Physical or manipulative operations will play a relatively insignificant part if we attempt to catch our science in its purely empirical stage. If occurrents are clearly given, they are so given without our inserting a finger into nature; if we are obliged to do something to nature before the occurrent is revealed, it is not obviously given but given only through the operation, and hence may not exist at all apart from the operation. Hence the data for a pure empirical science will be those occurrents which can be seen, heard, smelled, tasted, and touched, with only such manipulative processes as are required to make these immediate contacts possible.

But almost immediately certain more elaborate physical operations are introduced. These are associated with instrumental techniques on the one hand, and with experimental set-ups on the other. An instrument is a physical device for making clear an obscure occurrent. This is a satisfactory working definition, though it is inaccurate. More precisely an instrument is a device for producing in awareness a clear occurrent through whose content and operational derivation (i.e. the principles underlying the instrument) an obscure occurrent is given precision. It is because an occurrent is given obscurely that we employ an instrument. And when we employ an instrument we produce another occurrent, e.g. a deflection of a needle on a dial, an image seen through the eye-piece of a microscope, a line drawn on a moving chart, or the coincidence of the top of a column of mercury with a numbered line. Then, through this clearly given occurrent and a knowledge of the physical phenomena according to which the instrument was constructed, we explain the obscure occurrent. Operational techniques employing instruments are essentially different from the operational routes through which suppositional symbols are defined; the latter are inventive operations and always produce relatively obscure occurrents, but the former are operations of discovery and always produce occurrents which are in some respects more clearly given than the occurrents which we are trying to explain. Thus instruments are aids in the accumulation of data. But their use implies that we are dissatisfied with what nature herself reveals and consequently that we suspect her of possessing more information than she apparently does. Hence in forcing her to speak we are acknowledging the insufficiency of the strictly positivistic attitude toward nature; we are in essence recognizing that an adequate science must be more than descriptive. Thus already we can see the beginning of the passage into explanatory knowledge.

The same remarks apply to those manipulative techniques associated with experimentation. Experimentation may require instrumental devices, and in such cases all that we have said above applies. But experimentation may involve simply the placing of an occurrent in a situation in which it would not normally occur without the intervention of man. For example, we may separate things usually associated or associate things usually separated, or we may place an object in a foreign environment, or we may reproduce an object with one or more of its features modified. All of these would properly be called experimental situations and they would all presume something more than a purely descriptive interest in nature. For the implication of experimentation is that objects under different conditions may behave differently. We therefore recognize that nature is more than is obviously given, and we set about to determine what this increment of novelty is. This makes experimental situations into important aids in the accumulation of data. But in this stage of our analysis they must be looked at simply in that light. More specifically, any experiments which are performed by a purely descriptive science will be experiments for discovery rather than for verification. In this respect they will differ significantly from the usual scientific experiments. Normally an experiment is performed for the sake of verifying an hypothesis; hence there is a definite control of the set-up and an anticipation of the outcome. But in this case the experiments are of a random sort, involving a very minimum of direction and no advance knowledge of the outcome. They are best described as cases of toying with nature, poking or prodding her, turning her about, changing her habitat—all in the hope that we may learn something new about her.

One particularly interesting application of instrumental techniques is measurement. I think it is safe to say that a purely descriptive science will not employ quantitative methods. The justification of this assertion must be made later [1] where we shall give a more detailed discussion of quantitative methods. But here it may be stated that quantity is a relatively high abstraction; to note that qualities exhibit sufficient similarities to permit of classification and then arrangement in the class on the basis of an asymmetrical, transitive, and connected relation indicating difference in degree—to note this is to perform an abstractive operation of some difficulty. Hence we do not readily notice the quantitative aspects of things at all. A purely descriptive science is based merely upon presence or absence and not upon variation in degree. But the abstractive operation is one of the earliest of constructional operations; consequently measurement is almost certain to be introduced into a science as early as the character of the subject-matter permits. It is well

[1] Chapter XIV.

to note, however, that when a science takes this step it is becoming explanatory, for it is recognizing the legitimacy of talking about obscure occurrents, viz. quantities.

In summary, we may say of physical operations that a purely empirical science will not employ them; on the contrary it will confine itself to the delineation of what nature herself reveals, for these things alone are clearly given. But one soon feels that nature is more than she appears to be. Hence we introduce methods for determining what this supplementary factor is. These instrumental and experimental methods aim not to *create* entities but to *reveal* them; they are methods of discovery and not of invention. They are simply techniques for determining what nature is under conditions which vary slightly from the normal.

(*b*) The operations of selection are very important among the methods of descriptive science. The selective aspect of awareness, as was pointed out in chapter V, is one of its basic features. Mind is confronted with a realm of potential awareness which is vaguely perceived over a fairly wide area but precisely perceived only in a very limited area. The world of occurrents is complex in two different ways, both of which necessitate acts of selective attention. It is complex in a spatio-temporal way; every occurrent possesses relations, of more or less vague sort, to a more inclusive spatial and temporal situation, and to an included spatial or temporal situation. But the world of occurrents is complex also in an associative way; i.e. there are complex occurrents which are the togetherness of abstract occurrents. Every complex occurrent possesses relations, of a more or less vague sort, to its included associational elements and to the wider associational complex of which it may itself be an abstraction. Any occurrent is the terminus for these various types of relation to other occurrents. But it is characteristic of awareness that it finds its proper locus neither in the extensive nor in the minute spatial or temporal situations, and it finds itself able to grasp readily neither the high concretions nor the extreme abstractions. Thus it rests in intermediate occurrents. There is involved in every act of awareness an act of selective attention—an act of isolation[1] by which the occurrent is taken out of its structural setting and held up for consideration. Such an act easily becomes one of falsification if one loses sight of the character of the isolation. One must remember that isolation is neglect rather than negation, i.e. that the attempt is being made to grasp the unique rather than the structural content of the occurrent, and that, as a consequence, since structural features are often an important part of total content, the occurrent may appear quite different when taken in a more inclusive setting. The artificiality can be eliminated by a method which Lenzen[2] calls that of successive approximation by which the effects of idealizing the situation through isolation are progressively overcome by the step-by-step introduction of relevant features of the more extended situation. But if the isolative act is looked upon simply as neglect and one which does not involve a denial of the fact of associated and spatio-temporally related occurrents there is no essential danger. The neglected occurrent can always be recovered by another act of awareness directed upon the more inclusive situation.

There is perhaps some reason for insisting that this act of isolation is not properly an operation at all. In a sense it is not an act by which one passes from the awareness of one entity to the awareness of another, since, properly speaking, the entity with which the act begins does not enter into awareness at all. Or if it does it is only in an extremely vague manner. Furthermore, if it is an act involved in all awareness, there is no point in speaking of it as an operation; certainly it is not a transforming operation for if it were we should never be able to grasp the world as it is. All of this may be granted. The point is not important. But if the original occurrent enters into awareness with sufficient definiteness to be characterized as a somewhat, and if the act may be shown to be based upon actual relations in nature which, so to speak, guide the attention in its proper termini, then the act may be called an operation. To characterize it as an operation

[1] Cf. H. Levy, *The Universe of Science* (New York, 1933), chap. II; Hobson, *Domain of Natural Science*, pp. 38–9, 44–5; Cassirer, *Substance and Function*, p. 256; Poincaré, *Foundations of Science*, 362–68.
[2] *Physical Theory*, pp. 46–47.

of discovery or as an exploratory operation is to recognize precisely these facts.

The actual locus of this selective act cannot be determined in advance with any high degree of certainty. All that we can say is that description represents the beginning stage of knowledge and man naturally begins with man-sized objects. The obvious occurrents are the intermediate occurrents. They are not the widely extensive in space or time, nor are they the minute. They are not the high concretions nor the extreme abstractions. We are interested first in the things which enter into the spatial and temporal scope of our sensibility. This means that descriptive symbols are not about electrons, or atoms, nor are they about the world or the universe. We do not talk about the extremely great or the extremely small, for we do not know certainly that there are any such occurrents; hence the adequacy of our symbolic system demands that we exclude symbols for such entities. Similarly, we are first concerned with occurrents of moderate complexity—with things rather than with properties of things. High concretions, if they are recognized at all, are grasped through some less extensive complex of their properties, as when an individual is recognized by his walk. High abstractions, such as length, weight, color, and shape, if they are recognized at all, are observed only as elements of complexes, and not in isolation. Thus although abstract and concrete occurrents are *there*, as revealed by later awareness, they are not *obviously there* for descriptive awareness, and they are not selected out for attention. If, as is sometimes the case, relatively concrete objects are attended to before their elements, they are recognized as being obscure until their elements have been clearly discerned. In this way one might make a distinction between the temporal beginning and the logical beginning of a descriptive science. Temporally a science may begin anywhere—with high concretions or high abstractions, with highly extended or highly minute occurrents. What the scientist happens to discover first is largely a matter of accident. But in attempting to determine the character of the occurrents thus discerned one immediately becomes convinced of their relative obscurity as compared to the intermediate and middle-sized occurrents through which they must then be defined. If the data which are temporally first, are explicable only in terms of data which are temporally second, then the latter data must be logically prior to the former. Hence one should have done better to begin with these. It is the function of the selective act of attention not merely to limit the field of awareness but to direct it to those aspects of the total situation which are most clearly given and therefore consitute the material in terms of which the less clearly given aspects must be defined.

(c) The inventive operation employed in the derivation of the symbol is that of generalized characterization. We have already examined its nature.[1] It gives us the maximum directness possible in conjunction with economy and simplicity. By means of such symbols we talk about kinds of occurrents without specifying individual occurrents. This clearly involves a loss in the directness of reference; it means that science cannot be concerned with the individual but must talk simply about the type.[2] And it means that if the unique features of individuality are important for knowledge then science cannot lay claim to adequacy. The fact that the symbols of descriptive science are derived through generalized characterization introduces an element of conjecture into that science, for in the formation of symbols we always generalize beyond the scope of the observed situation and presume a reference to entities of the same kind not yet observed.[3] This danger can be minimized in a number of ways. One way is to supplement the characterizing symbols in each of their employments by proper names. Names enable us to determine the locus of a characterizing system. They perform the same function in a characterizing system that the symbol " Here you are " does as applied to a map—a function without which the map is utterly useless. Proper names help to retain directness in the reference of the characterizing symbols and we should expect descriptive systems to contain a large number of proper names. This is actually the case. Another way to retain this directness of reference is to employ such applicational symbols as " the," " a certain," " some," etc. The preponderance

[1] Chapter VI.
[2] Hobson, *op. cit.*, p. 29; Campbell, *What is Science?* p. 37.
[3] Hence general ideas are called " summational fictions " by Vaihinger (*op. cit.*, pp. 38–39; pp. 206–214).

of such symbols in descriptive science is a noteworthy fact. Again the closeness of the reference to occurrents can be retained by generalizing only from the concrete occurrents and not from the more obscurely given abstractions; thus there is less risk if we generalize from the clearly given particular to the obscurely given abstraction and then generalize this into a symbol. Finally the directness of the symbolic reference in descriptive symbols is indicated in the emphasis on extension as over against intension. Symbols are defined by pointing rather than through other symbols. Most of the symbols for terms take the form of " concrete " nouns rather than " abstract " nouns, i.e. symbols for occurents rather than symbols for universals or properties. So also the statements of structure, though they rarely occur, are in the form of correlations of occurrents rather than relations of universals. In all of these ways we endeavor to retain the directness in the reference of the symbol to its occurrent. There seems to be reason for saying that descriptive symbols exhibit a high degree of adequacy. At least they endeavor to retain the maximum of adequacy consistent with that type of symbol.

Adequacy of Correlational Symbols

A descriptive system is adequate as to its correlational symbols in so far as these refer to clearly given structures. This means that there are correlational symbols for all such structures and that they are all derived by the most direct sort of operational route. An interest in structure follows inevitably upon an interest in things; in fact if we neglect structure we can say almost nothing about things, and are limited simply to the symbolism of unique contents. What is ordinarily called description of an occurrent, i.e. listing its qualities, is really a listing of its relationships, for the designation of qualities is the symbolism of the more abstract occurrents with which the given occurrent is associated. Hence in order to know what qualities an occurrent possesses we must be able to select those occurrents with which it is related in a certain way; ordinary description is analysis of an associative complex into its abstract constituents. This necessarily involves pointing out relationships but they are of the more obvious sort since the abstractions themselves are relatively clearly given. There will be correlational symbols in a descriptive system, but they will in general be such as can be derived from clearly given structures through direct inspection. We shall therefore consider the question of the adequacy of correlational symbols with reference to the following points : (*a*) the type of correlational symbol which predominates, (*b*) the loose-knit character of the organization, (*c*) the presence of obvious gaps, and (*d*) the non-explanatory character of the knowledge.

(*a*) The correlational symbols which predominate in a descriptive system are those which are most directly derivable from the most obvious structures; these will be compatibility and non-implication, and temporal succession and spatial or temporal inclusions, in so far as these latter are expressive of occasional rather than repeated correlations. Correlational symbols are like characterizing symbols in that they involve the same loss in the uniqueness of the situations out of which they arise; thus even symbols for occasional structures will be obtainable from the occurrental realm by a somewhat indirect route. But symbols for universal correlations, viz. implications and co-implications, incompatibilities, causal laws, repeated inclusions,—all of these involve a greater leap, for in deriving them we assume a continuation of the structural relation in other situations, hoping that the uniformity of nature will guarantee such an assumption. Consequently we can say that a descriptive system which aims at a more direct relation between its symbols and occurrents will be loath to admit symbols for universal correlations ; repeated structures do not exist so obviously as do occasional structures.

An examination of actual cases of empirical science will show that this is the case. There are few laws, and no laws which are asserted with a high universality and necessity. Such laws as are found are usually statistical correlations, which presumably are closer to the facts than universal laws. The predominant types of propositions are the singular and the particular propositions of the Aristotelian logic, which assert relations of compatibility and non-implication. This is well illustrated in the employment of the case method in sociology. Furthermore, there are almost no causal laws. The cases which are clearly given are cases of *post hoc*, and from these we do not feel justified in inferring cases of *propter hoc*. The same is true with reference to the relations between propositions. Propositions are compatible with one another and non-implicative of one another, but there are few implicative, co-implicative, and incompatible relations. Just as structures are relatively more obscure then occurrents, so structures of structures are more obscure than structures. The structural aspect of the symbolic system is a reflection of the structural features of the realm of occurrents, but it is only the most obvious structure which is discerned.

It may also be pointed out that, just as a pure empirical science will not employ quantitative methods, so it will employ qualitative rather than quantitative correlations. But since quantity is more obscurely given than is quality, quantitative correlations will be given more obscurely than qualitative. Qualitative correlations are based upon mere presence or absence, but quantitative correlations indicate concomitant variations in degree. Such variations can usually be detected only by elaborate experimental methods which enable us to isolate features of complex occurrents and then to vary them while other features of the complex are maintained constant. This presupposes a rather highly developed ability to handle abstractions and is thus not a characteristic method of purely empirical science. However, since quantitative abstraction is one of the easiest of constructional operations to perform, it will be found in many apparently empirical sciences. But when it is found it indicates that the science concerned has already begun the transition to the explanatory form.

(*b*) The loose-knit character of the organization of descriptive symbols can be shown to follow from the preceding fact. If the structural relations which are symbolized in a descriptive system are merely those of occasional association, dissociation, succession, inclusion, etc., we must expect the body of symbols to be essentially aggregational rather than organic in character. This is because the only significant structural relations, i.e. the only relations which permit inference, are those of repeated structures. Consequently a body of descriptive symbols is hardly a system at all, not because it is internally inharmonious, but because the relations are mainly those of non-dependence. Relations of dependence determine a system, inharmonious relations disrupt a system, but relations of non-dependence make systemic considerations irrelevant. Structurally we know of any symbol only that it is compatible with certain symbols, and non-implicative with reference to certain others. These are not highly integrative relations and therefore do not permit inference; but since they do not permit inference they are not highly informative of the elements related. Hence in a descriptive system we can learn very little about an element by exploring its structural relations. To know with what occurrents an occurrent is occasionally associated or dissociated does not tell us much about the occurrent. The consistency is of a loose, aggregational sort, and we can speak of a " system " of such symbols only if the term can be applied to any totality in which there are no glaring inconsistencies.

With reference to the propositional elements of a descriptive science we shall be obliged to call attention to a similar looseness of structure. If propositions have no obvious relations of dependence they cannot be given logical structure. Hence a descriptive science cannot be deductively organized. The order of propositions in a descriptive science is the order of exposition rather than the order of argument. A descriptive science is a mere *collection* of true propositions bearing upon a certain subject-matter; as such the *order* of presentation makes no difference. There are no obvious incompatibilities, and some propostitions may be shown to be more closely related to a given proposition than others. But there will be little or no possibility of deriving one proposition from another; hence there will be no distinction between postulates and theorems.

It follows that a science in this stage will be continually undergoing revision, but each accretion or loss will involve a minimum of disruption to the body as a whole. Descriptive science proceeds in a piece-meal fashion. But the looseness of its organization prevents it ever suffering from a severe shock. Novel data may be absorbed and erroneous data may be rejected with a minimum of disturbance elsewhere in the system. It was because late nineteenth century physics had reached a relatively high degree of integration that it experienced such a tremendous upset at the advent of the theory of relativity. Empirical science does not undergo such experiences, but it suffers from a more serious defect. Because of the loose-knit character of the body of symbols there is no check upon observation; hence the system is almost certain to contain erroneous data. The principle of consistency cannot be employed as an observational test since the body of symbols has not yet taken on the force of an arbiter of experience, i.e. we do not know enough about the significant structural features of the world to predict what *must* be the case in a given situation. Therefore we are compelled to rely upon immediacy as the sole criterion of accuracy. The result is that we usually consider a descriptive science as temporary and approximate, exhibiting a high degree of flexibility, but progressively increasing in permanence, accuracy and rigidity through the process of absorbing new data and discharging erroneous data.

Another way of making clear this aggregational character of descriptive systems is to point out that in such systems we can neither *define* symbols nor *predict* happenings. Both definition and prediction rest upon knowledge of significant relations between symbols, and in the absence of such knowledge we can properly say that we do not know what the symbols mean and that we do not know what will happen in a situation in which one of the symbols participates. The case is not quite so bad for definition as it is for prediction. Though we are ignorant of the intension of the symbols we are not ignorant of their extension; hence although we do not know their structural contents we do know their unique contents. The absence of significant structural relations, then, does not imply that the symbols are meaningless; for they may always be defined through the denotative gesture; but it does indicate an important limitation of descriptive symbols and a consequent reason why such systems tend to pass over into explanatory systems. The route of extensional definition is only as precise as is the denotative gesture; but we cannot by this means alone distinguish between the various elements of an associative complex or between any one of its elements and the whole.¹ It will therefore be unsatisfactory for the definition of descriptive symbols if the occurrents we are endeavoring to symbolize happen to be abstractions which cannot be isolated from their complexes, or complexes which cannot be denotatively distinguished from their elements. But it is in just such cases that the intensional route will be more satisfactory, for we can then define the abstractions in terms of the concretions and the concretions in terms of the abstractions. These will be structural definitions and will be dependent upon a recognition of the corresponding structural features of the realm of occurrents. It follows that when definitions of this kind are found, the science has already recognized the necessity for talking about significant structures and has thereby ceased to be purely descriptive.

(c) A descriptive science is characterized by gaps. This is due to several features. In the first place it is due to the way in which the symbolic system is built up. Science may begin anywhere in the realm of the given, and its exploratory direction from this point is determined by a large number of unpredictable factors such as scientific genius, fortuitous discovery of scientific instruments, investigation arising out of practical needs, unexpected natural happenings such as earthquakes, eclipses, and pathological organic forms, negative results of experiments, and so on. Consequently there will almost certainly be some portions of the given which have not been adequately explored simply because the particular group of circumstances leading to such systematic exploration has not occurred. In the second place a descriptive science contains gaps because of the obvious inapplicability of the principle according to which its data are selected. It resolves to include only those occurrents which are obviously given. But the clearness of the given is a relative matter.² Hence if an occurrent is recognized as being present but not sufficiently clear to be symbolized there immediately occurs a gap in the symbolic system. All things which are more or less vaguely recognized by descriptive science determine

¹ Ogden and Richards, *Meaning of Meaning*, pp. 77-8.
² Feigl, *Theorie und Erfahrung in der Physik*, pp. 13-15.

gaps which must be filled in before the system can take on the proper integrative character. We feel these gaps whenever we ask the question " Why?" in descriptive science, but it is essential to the success of descriptive science, as we shall see immediately, that this question be rejected as illegitimate. In the third place descriptive science is characterized by gaps because of the absence of any principle by which occurrents may be anticipated. This is true only upon the stage of pure empiricism, for after a science has recognized even the most general sort of contentual and structural uniformity in nature the established body of symbols becomes regulative toward the future. But upon this early stage there is no such principle, and we are at the mercy of nature herself. At this point, given a present event, anything whatsoever may happen in the future and anything whatsoever may coexist with it. There are no anticipations, consequently no surprises and disappointments. But the method is very ineffective, for it does not permit that kind of observation which alone is successful, viz. observation which is guided by an advance knowledge of what is likely to occur, and it permits only the most random sort of experimentation.[1]

(d) A descriptive science does not explain. It is knowledge "that" but not knowledge "why." Every symbol in descriptive knowledge represents a brute occurrent—an occurrent which because it is unrelated to other occurrents might just as well have been other than it is. Explanation involves the establishment of relations of dependence. Such relations are not found in descriptive knowledge; hence nothing explains and nothing is explained. The discovery of a new fact is simply this and nothing more; it does not help in the understanding of an old fact. It might have been different, or it might never have been discovered at all; the descriptive system would have remained equally adequate. To explain an occurrent is to show how its symbol can be deduced from the symbol for another occurrent. Reason and consequence are found in implicans and implicate; hence a knowledge of the reason for things is dependent upon a knowledge of the structure of things. Since descriptive science is unwilling to recognize the significance of structure, it must

[1] Cf. Cohen, *Reason and Nature*, pp. 76–82.

abandon all hope of explaining. It must try to persuade men that to ask *why* is to ask an illegitimate question; it must insist upon the finality of description.
Nowhere is the instability of descriptive knowledge more evident than in this claim. The critics of positivism have been numerous in recent years and they have directed their arrows mainly at this point.[1] A discussion of the merits of these criticisms would take us too far afield. What we have already said in this chapter will indicate our position on the issue; what we are about to say as to the third feature of adequacy of descriptive systems will further clarify the problem.

Adequacy of Suppositional Symbols

A descriptive system is relatively inadequate as to its suppositional symbols. This means either that it contains few or no such symbols or that it contains some but recognizes their essential vagueness and consequently relegates them to a position of secondary importance. The characteristic feature of descriptive knowledge is its unwillingness to admit hypothetical and theoretical entities, idealizatons, constructs, and fictions into the body of science. The attitude of the descriptive scientist toward these entities may vary from one of a definite refusal to find a place for them, since they do not obviously exist and science can get along equally well without them,[2] to one in which they are recognized as having a certain heuristic value and even as possibly existing but they are then admitted only very reluctantly and only when they lead definitely to the discovery of new facts. In between these extremes will be the " as if " attitude which denies existence to such entities but admits them as explanatory notions. This is probably the status of Poincare's " conventions,"[3] Hobson's " ideal elements,"[4] Pearson's " constructs,"[5] Russell's

[1] Bavink, *op. cit.*, pp. 26-34; Max Planck, *Where is Science Going?* (New York, 1932), pp. 64-83; Meyerson, *De l'explication dans les sciences* (Paris, 1927), chaps. I, II, and *Identity and Reality*, pp. 384 *et seq*; Enriques, *op. cit.*, pp. 29-48; A. d'Abro, *Evolution of Scientific Thought* (New York, 1927), p. 380; Whitehead, *Concept of Nature*, pp. 44 *et seq.*, and *Adventures of Ideas*, pp. 159-65.
[2] Cf. Ritchie, *Scientific Method*, p. 38.
[3] *Foundations of Science*, pp. 123-6.
[4] *Domain of Natural Science*, p. 32.
[5] *Grammar of Science*, chap. II.

" fictions " and " logical constructions,"[1] Stace's " constructual existence,"[2] and other similar entities too numerous to mention. What is common to all of these attitudes is the recognition that suppositional symbols cannot refer to occurrents in the same direct manner that descriptive symbols do and that, as a consequence, (since the task of descriptive science is the most immediate portrayal of occurrents) they cannot be considered as a part of descriptive science.

It will be the task of the remainder of this chapter to show that descriptive science is compelled to admit the legitimacy of talking in terms of suppositional symbols. A purely empirical science is an impossibility. In recognizing this fact science opens the way for its own development in the direction of greater rationality and greater explanatory value. But the fact of the definability of suppositional symbols in terms of operational routes does not make such a development a movement toward greater artificiality. To talk about obscure occurrents is not to talk about unreal occurrents unless we talk about them in a careless manner. We shall here attempt to show (*a*) how the urge to be faithful to the clearly given necessarily involves an implicit recognition of the obscurely given, (*b*) how descriptive adequacy is strictly continuous with explanatory adequacy if the latter notion is properly understood, and (*c*) how the recognition of alternative routes of descriptive expansion implies the legitimacy of anticipation and hence of suppositional symbols.

(*a*) The demonstration of this point will require nothing new. We shall be obliged only to list the reasons thus far considered in this chapter why descriptive science, by the very nature of its claim to adequacy, finds itself obliged to talk about obscure occurrents and thus implicitly admits the legitimacy of employing symbols for things which are not clearly given. These points were as follows :

The use of physical or manipulative operations (instruments, experiments, mensuration) implies the recognition that nature as directly given is not the whole of nature. Nature as it is gives

[1] *Mysticism and Logic*, p. 128.
[2] *Theory of Knowledge and Existence*, chap. VII. The author here distinguishes between " constructual existence " and " factual existence."

hints of its character under alternative natural conditions. But these hints are simply obscure occurrents which we wish to talk about and which we endeavor to clarify.

The use of selective operations implies the same recognition. If awareness is selective, there is given a vague whole within which selection is made. But this whole is simply an obscure occurrent which we wish to talk about and which we endeavor to clarify.

The use of characterizing symbols has the same implication. Characterization is universal and involves a backward reference of the symbol to an alternative but similar occurrent which is given vaguely. But this alternative nature is simply an obscure occurrent which we wish to talk about and which we endeavor to clarify.

The use of correlational symbols has a similar conclusion. Correlational symbols, even when limited to specific correlational structures, contain a suggested reference to similar structures. But these similar structures are simply obscure occurrents which we wish to talk about and which we endeavor to clarify.

The use of quantitative symbols (correlational or otherwise) permits one to draw the same implication. Such symbols refer not to things but to aspects of things, and not to obvious structures but to hidden structures. But these aspects and hidden structures are simply obscure occurrents which we wish to talk about and which we endeavor to clarify.

The desire of the descriptive scientist to expand his system, fill in its gaps, and correct its errors involves a similar implication. A system could not be recognized as limited, or as containing gaps, or as erroneous unless there were hints as to the possibilities of extrapolating, interpolating, and correcting. But these additional and novel features are simply obscure occurrents which we wish to talk about and which we endeavor to clarify.

The acknowledged inadequacy of extensional definition in many cases implies this same recognition. If the denotative method is not always successful in locating an occurrent, this must be due to the fact that its essential content lies in its structural relations rather than its unique content. But these structural relations are simply obscure occurrents which we wish to talk about and which we endeavor to clarify.

The use of the very principle of clarity and obscurity itself implies the recognition of the fact that nature as directly given is not the whole of nature. Clarity and obscurity are relative matters; hence the obscure occurrents are simply those clear occurrents which were not considered sufficiently clear to come within the scope of formulation. But if our criterion had been applied with a greater liberality, then we should have wished to talk about them and to clarify them.[1]

(*b*) Description is not different in kind from explanation. In explanation we endeavor to determine the content of one symbol by deriving it from another symbol of which the former is an implicate. This is the strict conception of explanation, and it is limited to symbols which have this precise relation. But in the broader sense we explain any entity (not merely a symbol) by any other entity (not merely a symbol) when from the content of the former we find that we may determine (not merely through inferential derivation) the content of the latter. Now in this sense description is also explanation. Description is a process by which we derive the content of a symbol from the content of an occurrent. The derivation is not inferential, for there is no formal logic for the building up of symbols from occurrents, but there is recognized dependence of the symbol for its content upon the occurrent. Thus we should not be violating the meaning of terms seriously if we should say that we *explain* a descriptive symbol in terms of its occurrent.

But if description and explanation are both processes of deriving symbols from occurrents and determining their contents accordingly, what can be the distinction between them? There is a theoretical distinction which might be insisted upon. One might say that description is a process of deriving unique contents of symbols from unique contents of occurrents, and explanation is a process of deriving structural contents of symbols from structural contents of occurrents. Then description would be mere naming, and explanation would be interrelating the names. Then the task of descriptive science would be

[1] Stace, *op. cit.*, pp. 388–9.

precisely that of Adam when he was called upon to name the animals. We should remember that Adam was obliged to derive the symbols from the animals and must therefore have named them on the basis of what they " looked like." But the task of designating the various properties of the animals was not a part of Adam's job, for to recognize properties is to recognize the complexity of the object and the fact that it therefore *contains* aspects. However, " containing " is a structural relation and Adam, as a descriptive scientist, had no interest in it. So also with all of the other relations of the animals. Adam must therefore have been satisfied with merely pointing to the various animals, emitting the proper noise in the presence of each, and calling the result a descriptive science.

Clearly descriptive science is more than this. It certainly claims the right at least to point out *qualities*, and if qualities are simply symbols for other occurrents which are " contained " in things, then description insists upon the legitimacy of talking about this much structure. But if this admission is made, the possibility of a sharp line of demarkation between descriptive and explanation is abandoned. For the qualities of an occurrent are simply the *most obvious* of its structural associates, and the way is immediately opened for the admission of its *less obvious* associates—and we then have explanation. Hence a thorough-going positivism is confronted with this dilemma : Either science is confined to the deriving of symbols for unique contents of occurrents, in which case it is nothing more than the making of gestures to the accompaniment of the uttering of noises, or it is extended to the deriving of symbols for structural contents as well, in which case the positivist must admit that some structures are clearly given and others are obscurely given and consequently that description passes imperceptibly into explanation. The decision on this issue does not seem to me to be hard to make.

(*c*) The recognition of alternative routes in the expansion of a descriptive system also implies the admission of suppositional symbols. These alternative methods of proceeding in the development of a symbolic system are determined by the opposing demands of the growing system. On the one hand it must be as extensive as possible but on the other it must be as faithful as possible. There is the demand, which in many investigators is predominant, to retain the accuracy of the system by the refusal to admit any symbol which is not directly derivable from a clearly given occurrent. Awareness of symbols must follow upon awareness of occurrents, and we must progressively, if somewhat slowly, develop our symbolic system in a step-by-step manner, always recognizing its dependence upon the revelations of intuitive awareness. But there is another demand, which in certain investigators becomes the controlling urge, to speed up the expansion of the system even at the risk of error. Man soon recognizes his capacity to anticipate through fortunate guesses what intuitive awareness later reveals; he makes hypotheses, theories, and constructions in advance of the proper guarantee by nature. He then exhibits more interest in the intensional expansion than in extensional adequacy.

Clearly it is important to determine what there is which guides the guess of the investigator in this method of anticipation, for his guess may be supposed to be good or bad in proportion as it is guided or not guided. (This denies any special faculty of insight through which scientific hypotheses are discovered).[1] Since we have already considered this problem in our analysis of the methods for the derivation of suppositional symbols, we shall be brief here. It seems possible to say at least that the material upon which the guess works is the accumulated body of symbols with which the individual is acquainted, and that the requisite of the act of discovery is the consistent expansion of a limited system of symbols. Hence we can see what the investigator has to work upon in making his guesses and what is to be the guiding principle in the formulation of the guesses. This means that the scientist cannot make any guess he pleases; on the contrary, the range of guesses will be limited by the content and structure of the realm of clearly given occurrents. What we already know is always the most useful instrument for understanding what we do not yet know. Hence we are entitled to employ suppositional symbols in the exploration of the realm of the obscurely given,

[1] The possibility of such a faculty is according to Carmichael (*op. cit.*, pp. 8, 18), one of the reasons for the failure of a logic of discovery.

provided we recognize the principles guiding the definition of such symbols. Random and unguided guesses are expressed in vague and ill-defined suppositional symbols, but systematized and controlled guesses exhibit themselves in operationally defined suppositional symbols. Consequently it is not a question of the legitimacy or illegitimacy of making guesses—one cannot, perhaps, avoid making guesses—but a problem of employing the proper operational techniques in the advance formulation of the guesses. But in admitting the propriety of guessing in science we have abandoned more or less completely the claim that a descriptive science may be sufficient. Hence we are already passing over into considerations pertaining to explanatory science.

The special feature of this problem, to which we shall now refer in transition to the subject-matter of the next chapter, is as follows : Since the established symbolic system is the foundation for the determination of the content of the suppositional symbol, the more highly developed that system is, both as to inclusive adequacy and as to high integration, the greater the warrant will be for its expansion into the realm of the conjectural. Hence the method of anticipation becomes increasingly applicable as the symbolic system takes on a higher and higher degree of development. This is due to the greater wealth of material upon which to base the suppositional symbol and to the greater degree of consistency within the symbolic system, which permits of greater definiteness in the formulation of the operational route.

CHAPTER XIII
EXPLANATION

In our discussion of explanatory knowledge we may follow again the outline of the preceding chapter. There we pointed out the extent to which descriptive knowledge satisfies the general claims of symbolic adequacy. With reference to descriptive symbols this form of knowledge was seen to be highly adequate. With reference to correlational symbols it appeared less highly adequate, since it recognized vaguely the existence of types of structure which it could not symbolize. With reference to suppositional symbols it proved least adequate, since it admitted these symbols only very reluctantly, in spite of the fact that the very implications of the descriptive method are such as to demand their inclusion. We may now raise the same questions concerning explanatory knowledge. What is its adequacy (1) as to descriptive symbols, (2) as to correlational symbols, and (3) as to suppositional symbols.

But here again we must anticipate a difficulty which we noted in descriptive knowledge. *Pure* explanatory sciences are as hard to find as are *pure* descriptive sciences. In fact the character of the distinction between them is such that there cannot be pure sciences of either kind. This means that every empirical science exhibits anticipations of rationality and every rational science retains its empirical reference. We can determine the properties of a purely rational science only by extrapolation from the series of increasingly rational sciences in the direction of rationality.

However, the situation is somewhat more complex in the treatment of rational science than in the consideration of empirical science. Empirical science starts, at least ideally, *de novo* from the realm of occurrents. Hence we need not know what rational science is in order to understand empirical science. But rational science starts from and arises out of descriptive science. Hence we can understand neither its structure nor its content unless we know something of the character of descriptive science. The reason for this was suggested in the closing paragraph of the last chapter. Explanatory science arises precisely because of the recognized inadequacies of descriptive science, but, strangely enough, it must build upon the foundations laid by descriptive science. Consequently the content and structure of explanatory science are founded upon the content and structure of descriptive science, but not simply upon these. Explanatory science arises when there is an acknowledged admission of the inadequacy of descriptive science in its treatment of obscure occurrents, and when there is also an acknowledged admission of the necessity of taking obscure occurrents into serious consideration. Hence the problem of explanatory science is that of determining the nature of what we do not yet know clearly about a given subject-matter upon the bases of what we know clearly. What we are capable of knowing is therefore founded upon the extent of the descriptive system and upon the degree of integration. As we learn more about the realm we find ourselves in a position which enables us to venture forth beyond the obvious matter of the system with a feeling of greater security. The guesses of a mere novice in a field of investigation have a very low degree of probability as measured against the insights of a well informed student. The content and structure of the accumulated knowledge become the pattern for the expanded system. To the extent to which the descriptive system is highly integrated, relatively extensive, and without significant gaps, the movement to the suppositional symbol is guaranteed. If the given system is pervaded by a structural feature there is reason to suppose that this structural feature extends beyond the limits of the obvious system and thus brings into the system those further elements which, perhaps, have already been vaguely foreseen. Hence one might say that an explanatory science is any science in which we talk in a relatively precise manner about occurrents which in descriptive science are given only obscurely. Then one must recognize that the possibility of talking about these occurrents precisely is dependent upon the awareness of certain other occurrents, themselves clearly given in the descriptive science.

Adequacy of Descriptive Symbols

It is probably correct to say that highly developed rational sciences are relatively inadequate as to their descriptive symbols, and we shall attempt immediately to point out what this involves. But it is erroneous to say—as is implied in the characterization of explanatory sciences as non-existential, ideal, and *a priori*—that they do not have and perhaps never did have any relation to the realm of occurrents. For example, it is incorrect to say that because geometry talks about ideal and perfect figures it therefore does not refer to actual objects at all. To say this is to suppose that when a science introduces suppositional symbols it abandons all descriptive symbols. This is not the case, though the change in emphasis leads one to believe that it is so. When a descriptive science has taken on a relatively high degree of integration the interest of the investigator shifts from the descriptive symbols to the correlational and suppositional symbols. In the terminology to be introduced in this chapter the interest of the scientist passes from the *theorems* to the *postulates*.[1] This does not mean that a deductive system contains postulates but no theorems; it means that the theorems are contained implicitly rather than explicitly; we do not need to mention the theorems because we all know that they are true, and that they can be derived from the postulates. Consequently when geometry talks about ideal figures rather than actual figures it has not abandoned the descriptive reference; it has simply recognized that the properties of actual figures need not be talked about because everyone knows what these properties are and that they can all be deduced from the postulates about ideal figures. Thus one should say that a system of geometry about ideal entities is itself a postulate system from which one can deduce (if a postulate stating the relation of ideal entities to actual entities be introduced) all of the propositions about actual entities. For this reason when we say that an explanatory science is inadequate as to descriptive symbols we should be clear as to what it involves.

But there is a sense in which explanatory knowledge is *less interested* in the adequacy of its descriptive symbols, and we must turn immediately to this. Following the analysis of the

[1] Cf. Carmichael, *op. cit.*, p. 49; Rueff, *op. cit.*, chap. V.

preceding chapter we may say that explanatory science relies less than descriptive science upon (*a*) physical or manipulative operations, (*b*) selective operations, and (*c*) direct inventive operations in the formation of symbols. Since these features have already been discussed as applicable to descriptive knowledge, they may be considered somewhat briefly.

(*a*) The reliance upon manipulative operations will depend upon the degree to which the explanatory science has developed. The employment of such operations, as we have seen, is an indication of a science already passing out of the purely empirical stage and beginning to take on certain features of rationality. It indicates an increased interest in what nature might be, or is obscurely, as over against what it is obviously. But the use of physical operations implies that the occurrents about which we wish to learn are capable of being given more clearly if the proper physical situation is produced; the use of experiments, instruments, and measuring techniques, brings about just such situations. We are no longer obliged merely to listen to nature; we may now proceed to ask her questions. By means of suppositional symbols we are able to anticipate nature and subject her to such conditions as will make her reveal her secrets. We perform guided experiments which may be crucial in enabling us to decide whether a certain kind of occurrent exists in nature. We reproduce nature in the laboratory under controlled conditions, augmenting minute phenomena, diminishing gigantic phenomena, isolating phenomena from their usual associates in nature, and so on. This is the essential value of the suppositional method, and is completely disregarded by those misguided Baconians who see in hypotheses nothing but weapons whose danger to the user more than offsets their efficacy.

But it can be seen, I think, that as a science takes on a higher and higher degree of rationality it tends to abandon manipulative methods. As purely empirical it is interested only in the occurrents which are clearly given without any manipulative processes. As an embryonic rational science it is interested also in those occurrents which can be clearly given if the proper physical situation can be produced. As a developing rational science it becomes interested also in those occurrents which could be clearly given only if a physical situation not yet capable of being constructed could be produced. As a highly developed and mature rational science it becomes interested in those occurrents which contain an inherent obscurity because of the theoretical impossibility of isolating them from other more clearly given occurrents with which they are always associated. For example, a purely empirical science might be interested in the obvious features of an object, say its shape, color, odor, or weight. An embryonic rational science might be interested in its character as revealed by cutting it into parts, pouring acid upon it, or heating it. A developing rational science might be interested in its atomic or molecular structure. A highly developed and mature rational science might be interested in the general fact of shape, which is obscure because abstract and incapable of being isolated from the particular shape with which it is always associated. From this it follows that as a science becomes concerned with high abstractions it has less reason to rely upon manipulative methods. If abstractions always occur in association with concretions, manipulative techniques will be effective only in so far as through them we may produce the greatest possible variety of concretions and thus see more and more of the implicans of the abstractions. (For example, we learn something novel about life every time we discover a new living form). So also manipulative techniques will be of no use in studying high concretions for the concretions must always occur in association with abstractions, and there are no isolative operations by which they may be singled out. Manipulative techniques may be employed in determining the character of wholes and parts only to the extent to which the respective processes of synthesis and analysis are valid. Such techniques may never be employed in studying the past. The result is that as a science takes on a rationalistic interest it tends to discard physical techniques and begins to take on " mental " techniques, i.e. it recognizes that many of the most important of the obscure occurrents are really given in the situation but must be discerned from their associates by a difficult operation of selection. To a consideration of this we now turn.

(*b*) Selection is an operation by which whatever is given obscurely takes on clarity through direction of attention; it is an act by which awareness moves from one portion of the field of the given to another portion associated with it or related to it spatially or temporally. For our purposes one feature of this selective act may be called to attention, viz. that it becomes habitual and unconscious as one continues to perform it. From this an important result follows. One is no longer able to say whether he selects the obscure occurrent from the background of the more clearly given occurrent and then defines the symbol for it in terms of what he discovers to be its nature, or whether he performs an operation upon the symbol for the clearly given occurrent and then defines the symbol for the obscurely given occurrent in terms of this operational derivation. More briefly and more specifically, one is unable to say whether he defines symbols for abstract occurrents by isolating these occurrents from their more concrete associates and then determining the contents of the former by direct inspection, or whether he defines such symbols in advance of the awareness of the proper occurrents through an abstractive operation performed upon the symbol for the more concrete occurrent. It may appear that there is no essential difference between these two methods. But there is an important difference. According to the former the symbol for the abstraction is defined by examination of the abstract occurrent, but according to the latter it is defined in advance of the examination of the abstract occurrent by inference from its concrete associate. The difference is essentially that between the method of interpretation and the method of anticipation.

Now, as was pointed out, the fact of habituation to the performance of operations leads to an obliteration of the distinction between these two methods. But it is important to decide which of the two methods is more basic and therefore to be taken as the one to which the other is reducible. The answer to this question determines one's attitude on the more general question of the status of each of the two types of science. If one insists that all defining operations through which we apparently anticipate nature are reducible to selective operations through which we explore nature, then rational science tends to become existential, real, and *a posteriori*. But if one insists that all selective operations through which we apparently explore nature are reducible to defining operations through which we anticipate nature, then rational science tends to become nonexistential, ideal and *a priori*. According to the former all things that science talks about exist and are discovered; according to the latter most of the things that science talks about do not exist and are inventions.

The point of view which I have taken on this issue should now be clear. We can say neither that all of the things talked about by science exist, nor that most of them do not exist. What we should say is that if an occurrent is obscure and does not obviously exist we cannot talk about it until we have symbolized it precisely. Now if we can select the occurrent from its background and thus become clearly aware of it, this is the proper way in which to define its symbol. But if we cannot do this we are compelled to resort to an indirect method; we define the symbol in terms of the more complex situation of which the occurrent is a part and in terms of the structural relation of the occurrent in question to this situation. This enables us to determine in advance of awareness what content the occurrent must possess—if nature exhibits the structural uniformity which we attribute to her. Whether the occurrent in question exists cannot be determined in any general way, since it depends upon the nature of the operation (whether it is determinate or indeterminate) and the operation's own existential claim (whether or not it refers to an occurrential structure)— which is in turn dependent upon whether the operation is public and repeatable. Consequently it is impossible to say anything in general about the existence or non-existence of the entities of rational science. What we should say is that the task of science (both empirical and rational) is the discovery of occurrents; but this task does not in any way preclude the possibility of talking about occurrents before we discover them, nor does such advance awareness make the entities inventions and thus preclude the possibility of their being discoveries as well.

(c) Since the emphasis of explanatory science is upon the obscurely given rather than the clearly given, there will be few directly inventive operations for the derivation of symbols. In other words in explanatory science we are not primarily concerned with naming the given. If we do employ definitive operations at all we must use them, as we shall see, through the structural features of the system. A symbol takes on meaning not so much through what it refers to as through its relations to other symbols. Hence we need not reconsider the operation of generalization.

Adequacy of Correlational Symbols

An explanatory system puts an increased emphasis upon correlations, particularly upon those correlations which are based on repeated structures. We pass on, in explanatory science, to consider those structures which were given only obscurely in descriptive science. The adequacy of explanatory knowledge from the point of view of its correlational symbols will be determined with reference to the following points : (a) the preponderance of highly determinate correlations, (b) the tightly-knit character of the organizaton, (c) the absence of gaps, and (d) the explanatory, i.e. deductive, character of the organization.

(a) Instead of relations of compatibility, non-implication, and independence, there will be relations of implication and co-implication. The predominant type of proposition will be the universal proposition of the Aristotelian logic. There will be laws, and they will partake of a relatively high degree of universality and necessity. Statistical correlations will be replaced by unexceptionable connections. Wherever possible quantitative correlations will be employed, since the science has already passed through that stage of developing empiricism in which isolational methods permit the ascertainment of concomitant variations ; such quantitative methods permit an extraordinarily high degree of integration, as will be indicated in the following chapter. Propositions will have the same highly determinate relations as do the simpler symbols ; thus they will in general be related as implicates or co-implicates or incompatibles rather than as mere compatibles and non-implicates.

(*b*) The result is that the symbolic elements take on a high degree of integration.[1] This integration is expressed in deductive structure, of which we shall make an analysis presently. The body of symbols has ceased to be aggregational and has become organic or organized. Elements are dependent upon one another and may be inferred from one another. To know the place of an element in the system is to know something highly informative as to its character. Hence in such an integrative system we can learn much about an element by exploring its structural relations. To know with what occurrent a given occurrent is *occasionally* associated does not tell us much about it, but to know that to which it is *inevitably* and *universally* related tells us many important things about it and about the general structure of nature. We now have a *system* in the precise sense of the word.

It follows that explanatory science has a relatively high instability as compared with descriptive science. The latter is continually being increased and corrected, but each addition and revision involves a minimum of disruption to the structure as a whole. Explanatory science is different. It must add to itself, take away from itself, or correct itself with the greatest caution, for any addition, subtraction or transformation may disrupt the organization. The organization is unstable because so complex. Though empirical science is continually changing, its growth is uniform and progressive; explanatory science changes little and its growth is usually through severe shocks which often nearly destroy the system, e.g. non-Euclidean geometry, non-Aristotelian logic, quantum mechanics, special and general theories of relativity. Frequently such shocks involve a more or less complete revision of the system.

The transition from descriptive to explanatory science indicates how this sensitivity to shocks increases with the development of the integrative character of the system. An explanatory science takes on an intensional emphasis to replace the extensional emphasis of a descriptive science. A symbol in a descriptive science is therefore defined primarily in terms of

[1] Cf. Cohen, *op. cit.*, pp. 106–114; Hobson, *op. cit.*, chap. IV.

its extension and only secondarily in terms of its intension. But as the symbolic system increases by the addition of data, new structural relations are discovered, and the question inevitably arises as to whether we ought to retain the definition of a symbolic element in terms of its extension or to redefine it in terms of the newly discovered structure. So, later, when still more data are collected, the question arises as to whether the symbol ought to be defined in terms of its old intension or in terms of the newly discovered intensional relations. What actually happens is that the symbol is redefined from period to period as the knowledge of its structural relations increases. This is well illustrated in the historical development of the definition of a chemical substance from its physical properties, through its general chemical properties, then in terms of certain specific chemical properties, and finally in terms of molecular structure and atomic weight. Each new structure supersedes the others as a defining property.

But this cannot go on indefinitely. A limit is being approached by virtue of the belief that each new definition is more nearly correct than the preceding; hence there is less likelihood of a later definition being refuted or corrected by experience. Consequently a highly developed explanatory science is not so completely exposed to such shocks. In fact one may say that an explanatory science, by virtue of its high integration, becomes regulative of experience so that in a sense its own disruption is made impossible by *fiat*. This is what is meant by saying that definitions are verbal and incapable of being refuted by experience. A definition is the statement of the relation of inclusion between an associative complex and two or more of its elements. In general such a statement may be used either for the determination of the character of the complex from the characters of the elements or for the determination of the characters of the elements from the character of the complex. Usually, however, definition is limited to the former; this is based upon the presumption that elements are always more clearly given than complexes. But granting this presumption and this more limited application of the notion of definition, one may see how a definition once established becomes regulative of future experience. If gold is defined as that substance which melts at $1075°$[1], then we have determined in advance that we shall never find a case of gold which melts at any other temperature; for if such a thing should be found, we should deny that it can properly be called gold. In the first place, if gold is identifiable *only* in terms of this property, then we should have no means of characterizing a newly discovered substance as gold, and we should therefore properly deny that it *is* gold. In the second place, if gold is identifiable also in terms of some other property, say specific gravity, then the definition should have so stated; and since in this case gold is identifiable in terms of both melting point and specific gravity, the new substance would still not be gold, because it would lack one of the two essential properties. This anticipatory feature of a symbolic system is well recognized by science. As a science takes on a higher integration it becomes more definitely regulative, since its correlational symbols assert connections with a higher degree of necessity. Through significant intensional relations each symbol receives substantiation from each other. Hence empirical exceptions are met not by a revision of the conceptual system, as is the case with descriptive knowledge, but by denying the applicability of the symbolic formulation to the situation. Thus symbols take on a necessary, *a priori* character and the total system of symbols is protected from reversals. This seems to me to be the truth in the position of C. I. Lewis.[2]

(*c*) An explanatory differs from a descriptive science in having filled in the gaps which are characteristic of the latter. These gaps, we found, are due in a descriptive science to its unpredictable origin and to its dependence for development upon all sorts of fortuitous circumstances. But by the time a science has become explanatory the realm of its subject-matter has been more or less completely, if somewhat unsystematically, explored. Putting it as simply as possible, explanatory science does not contain gaps, merely because we know so much more about the range and character of the data. Gaps are also due to the unwillingness on the part of the descriptive scientist to admit

[1] Cf. Chapter IV, p. 75.
[2] *Mind and the World Order*, chap. VIII; Cf. also Cohen, *Reason and Nature*, pp. 99–106, 142–3.

obscure occurrents into the realm of legitimate subject-matter; such occurrents are recognized as present, but they are not recognized as things to be talked about and included in the system. Explanatory science, however, admits frankly that any occurrent which is not *too* obscure may be properly admitted and talked about provided we symbolize it with caution and through adequate operational definition. Finally, gaps may be due to the lack of a principle according to which the data may be anticipated. But if a science has taken on a highly integrative character, it has, as we have just seen, precisely this directive character toward the realm of data. Thus gaps are filled in because we know in advance what to look for and where to look.

These gaps may be of several kinds. Dissociative complexes which do not possess significant spatial or temporal organization are obstacles to integration. These are usually inserted into the symbolic system by discovering if possible an abstraction of which the elements of the complex are concretions; in this way the fact of difference, which is an apparent impediment to rationality, is replaced by similarity which yields to rational formulation. This is what we always do when we classify. One of the best examples of it is in the principle of serial organization which permits occurrents exhibiting a high degree of difference to be united by a common abstraction through the insertion of a continuous series of elements gradually changing from the one to the other. For example, a circle may be shown to resemble a triangle if we circumscribe the circle about the triangle and then proceed to double repeatedly the number of sides of the triangle and the other inscribed figures; in this way the continuity of properties is revealed. Other gaps may be filled in simply by definition. If we have given a repeated association of occurrents the integrative character of the whole may be increased by merely naming it. Other gaps are not filled in but eliminated as unreal, as in the case of perpetual motion machines and of the duplication of the cube. Still other gaps, of a more common sort, are filled in simply through the symbolism of the obscure elements through which they may be explained; this is the status of all hypotheses. In these ways the highly integrative character of explanatory knowledge is attained by removing the obstacles to deductive organization, discovering abstractions and concretions, eliminating the irrationals and surds, explaining away unreal difficulties, and inserting suppositional entities where they are required.

(d) Explanatory knowledge does not merely describe; it also explains. It becomes legitimate to ask the question, Why? There are no brute facts. Nothing in the system could be otherwise without involving a more or less complete disruption of the system. Thus with reference to every derived proposition in the system we can see *that* it is so (by empirical corroboration) and *why* it is so (by deduction from underived propositions); and with reference to every underived proposition, although we cannot see *that* it is so (since empirical corroboration is dependent upon the awareness of obscure occurrents), we can nevertheless see *why* it is so (by virtue of the truth of its implicates). The more complete examination of this notion involves a consideration of the structure of deductive systems,[1] to which we now turn.

Deductive Systems

Explanation is possible by virtue of the structural relation of concretion to abstraction. We explain a symbol when we reveal another symbol of which the given symbol is an implicate. Thus explanation is possible in a symbolic system wherever it can be shown that two symbols are related as implicans to implicate. Upon the basis of such a relation it is possible to divide the elements of integrated symbolic systems into two groups, the derived elements and the underived elements, or (since a deductive system is possible only with reference to propositional

[1] Since the analysis of abstract deductive systems has become a highly specialized problem in recent years, I shall not raise considerations of this kind. The thesis which I am endeavoring to defend can be made clear without reference to them. Some of the references upon the less abstract aspects of deductive systems are as follows: Cohen, *op. cit.*, pp. 106–14; Keyser, *Mathematical Philosophy* (New York, 1922), chaps. II–VIII, also *Thinking About Thinking*, chaps. II–IV; Carmichael, *op. cit.*, chaps. III, V, VI; C. I. Lewis, and C. H. Langford, *Symbolic Logic* (New York, 1932), chap. XI, sec. 1; P. Weiss, "Nature of Systems," *Monist*, Vol. XXXIX, pp. 281 ff., 440 ff.; J. W. Young, *Fundamental Concepts of Algebra and Geometry* (New York, 1927), chaps. IV, V.

elements) derived propositions and underived propositions. The underived propostitions are those elements which are implicans but not implicates, the derived propositions are those elements which are implicates of the underived propositions. Some of the derived propositions may be implicans of other of the derived propositions, but all must be implicates of the underived propositions. This is the usual distinction between postulates (as including axioms and definitions) and theorems. The postulate system is the totality of underived propositions; the theorem system is the totality of derived propositions.

For our purposes the important feature of this organization is the way in which a deductive, explanatory system is related to the descriptive system out of which it arose and which it aims to explain. Put crudely, the explanatory system contains the descriptive. More precisely, the theorems of the explanatory system are exactly the propositions of the descriptive system. These are all about clearly given occurrents, and their truths are determined by extensional reference. The postulates of the explanatory system are those propositions about obscure occurrents which the descriptive scientist either refused to admit into his system or else tolerated only because they seemed to have a certain heuristic value. Since the theorems may be derived from the postulates, we can now see what was meant above by saying that the theorems are true both by empirical corroboration and by deduction, while the postulates are true not by direct empirical corroboration but by their capacity to imply the descriptive propositions.

The problem of the transition from descriptive science to explanatory science is the problem of the discovery (or invention) of postulates, and this problem is precisely the same as the inductive problem of the discovery (or invention) of hypotheses. Thus when the mathematician is searching for a postulate system from which he may deduce all of the theorems about the cardinal numbers he is doing the same sort of thing which the natural scientist does when he looks for entities like molecules from which he may deduce all of the descriptive propositions about heat, expansion, vapor tension, solubility, etc. Furthermore, the problem of whether these postulate systems are invented or discovered must be solved as the similar problems of suppositional symbols in general have been solved. If one believes that the obscure entities really exist and can be discerned through a proper selective act from their more clearly given background, then he will insist that they are discovered; but if one believes that the obscure entities may not exist and hence must be defined in advance of awareness through operational routes from the more clearly given occurrents, then he will be inclined to feel that they have been invented. But regardless of how this problem is solved, the method for the ascertainment of postulates must be based upon a recognition of the structural relation of the complete postulate system to the descriptive system which is to be explained. Hence we must examine this relation more closely.

We have suggested that the two main kinds of element in a deductive system are the underived propositions and the derived propositions. The former are illustrated by the theorems of geometry, the empirical laws of the natural sciences, the statistical generalizations of the social sciences, and all particular and singular propositions. The latter are illustrated by the postulates, axioms, and definitions of geometry, the principles of the natural sciences, and the methodological assumptions of all of the sciences. But this latter group is capable of being subdivided into those propositions which are the postulates proper, in that they are assertions about the subject-matter of the system, and those propositions which are the axioms or principles of procedure and thus justify the inference from the postulates to the theorems. The principles of procedure are not usually stated as part of the group of underived propositions, but are taken as implicit. Sometimes, when they take on a relatively specific character, they are inserted along with the postulates, as in the case of the axioms of quantity in the Euclidean system of geometry. They are the principles of logic which justify the determination of the content of the postulational system upon the basis of the descriptive system and the inference from the postulational system to the descriptive system. More specifically they are the principles which justify the definition of abstractions in terms of concretions and concretions in terms of abstractions, wholes in terms of parts and parts in terms of wholes, effects in terms of causes and causes in terms of effects. Conversely they are the principles which justify explaining concretions in terms of abstractions and abstractions in terms of concretions, parts in terms of wholes and wholes in terms of parts, causes in terms of effects and effects in terms of causes. It was the task of chapter IX to make a detailed analysis of one of these relations, viz. that expressed in the structure abstract-concrete. It was also suggested in that chapter that analytic, synthetic, and causal operations constituted alternative and supplementary methods. We have now seen the role which these principles play in the constitution of a deductive system.

With reference to the postulates proper there is another important distinction. Postulates constitute the total body of propositions which make affirmations about obscure entities, hypotheses, theories, constructs, fictions, high abstractions and concretions, highly extended and minute, past and future, etc. These will be divisible into two groups which are called by Campbell the "hypothesis" and the "dictionary."[1] Expressed in our terminology the former group consists of all propositions which interrelate suppositional symbols amongst themselves, while the latter group consists of all propositions which interrelate the suppositional symbols with the descriptive symbols. Then the hypothesis constitutes a group of propositions which assert the structural interrelations of high abstractions and concretions, highly extended and minute occurrents, past and future events, etc. And the dictionary constitutes a group of propositions which give the rules for "translating" the hypothetical system into the descriptive system. Campbell's example[2] is incorrectly analyzed according to his own definitions of these two notions. What he has included in his hypothesis

---

[1] Campbell, *Physics, the Elements*. p. 122. Cf. also Lenzen, *op. cit.*, pp. 39–42. I believe that Lenzen's "theoretical physics" corresponds to Campbell's "definition" and Lenzen's "mathematical physics" to Campbell's "hypothesis." In Campbell's other discussion of this topic (*What is Science?*, chap. V) he uses the term "theory" to include both "dictionary" and "hypotheses." Cf. also Poincaré, *op. cit.*, pp. 126, 335–6. Poincaré points out that an empirical law may be elevated into a principle or "convention;" this seems to coincide with Campbell's "hypothesis." But the other empirical law which remains for Poincaré does not coincide with Campbell's "dictionary." I have never been able to ascertain just what the function of this remaining element is.

[2] *Op. cit.*, pp. 123–4.

K

should be included, but part of what he has included in his dictionary should have been included in the hypothesis. The complete hypothesis should consist of a set of algebraic equations expressing functional relations between certain quantities designated by letters having specified variability or constancy, range, etc. And the complete dictionary should consist of the totality of assertions defining R as the resistance of a piece of metal, T as the temperature, and a certain functional relation as the statement that the ratio of the resistance of a piece of pure metal to its absolute temperature is a constant. The former group interrelates quantitative symbols among themselves and the latter group relates quantitative symbols with symbols for directly given occurrents. The statements in the hypothesis, as Campbell points out, appear as arbitrary assumptions apart from the dictionary. But his use of the term " dictionary " does not seem fortunate to me, for it suggests that the relation between the suppositional symbol and the descriptive symbol is one of equivalence. Let us attempt to see what the character of this " dictionary " relationship must be.

There is almost certain to be confusion if we do not use terms very precisely at this point. This is due to the fact that the total postulate set implies the theorems and thus is concrete with reference to them, while the suppositional symbol itself may be an abstraction from the symbols for clearly given occurrents and hence be implied by the theorems. Specifically we must distinguish between (*a*) the suppositional symbol itself, and (*b*) the total postulate set stating all of the truths about the suppositional symbol. The suppositional symbol will be a word such as " molecule," " electron," " perfect lever," " point," " ether," or an algebraic symbol for a quantity, which designates any obscurely given occurrent operationally defined and employed as a symbol of explanation. The total postulate set is a group of propositions asserting all of the structural relations of the suppositional symbol to other suppositional symbols and to symbols for clearly given occurrents. This complex must be concrete with reference to the body of theorems since it implies them. But this total postulate set must be divisible into three groups rather than two as Campbell maintains. These will be

(1) the *dictionary* or the totality of those postulates which state the operational connection of the suppositional symbol with the data, (2) the *construct* or the totality of those postulates which state the content of the suppositional symbol so far as this is inferrible from the descriptive symbols through the operational route, and (3) the *hypothesis* (which includes the construct) or the totality of those postulates which state the content of the suppositional symbol in its capacity to explain the descriptive propositions. Let us discuss each of these.

The *dictionary* states how the suppositional symbol was derived and therefore, in general, how it refers back through the total postulate set to the descriptive propositions. If the postulate set is to imply the descriptive system, there must be at least one proposition in this set which asserts a relation between the suppositional symbol and the symbol for a clearly given occurrent; otherwise there would be no connection between the postulate set and the theorems. For example, given an abstract postulate set such as that offered by Lewis[1] for the Boole-Schroeder logic, the propositional interpretation can be shown to follow from this set only if there are introduced further postulates which state that the various elements represent propositions and the various relations represent propositional relations. Such a body of propositions would constitute the dictionary of the total postulate system. But this is only one type of dictionary relation, since it states that the suppositional symbol is related to the descriptive symbols by abstraction. To be sure, abstraction is one of the most important of such relations ; as we shall see in the last chapter, quantitative theories are all of this general type. However, we have seen that a suppositional symbol may also be more concrete than the descriptive symbol from which it was derived, or it may be an associate. Thus there are dictionary relations corresponding to the three methods for the formation of constructs, viz. association, abstraction and concretion. But there are still others which we have not discussed in detail. Some of these are analysis and synthesis, which state respectively that the suppositional symbol refers to a spatial or temporal part of that which is referred to

[1] *Survey of Symbolic Logic*, p. 119.

by the descriptive symbol, or to a spatial or temporal whole of which that referred to by the descriptive symbol is a part. Others of the constructional operations are causal inferences by which we define the content of a suppositional symbol for the past or for the future in terms of a descriptive symbol for the present. These are types of dictionary relations since they enable us to determine the content of the suppositonal symbol from the content of the descriptive symbol and also permit us to infer from the total postulate set, as later determined through the hypothetical method, to the total body of descriptive propositions.

The brevity with which the dictionary relation is here being discussed, should not lead one to conclude that it is an unimportant relation in the problem of the logical structure of science. On the contrary, the topic may be treated briefly precisely because the entire book has been more or less directly about it. The solution to the problem of dictionary relations is the solution to *the problem of the relations of perceived entities to the entities of the scientific world*, i.e. the problem of " bridging the gulf between the world of sense and the word of physics,"[1] It seems safe to say that this has been the outstanding epistemological problem from the time of Locke down to the present; certainly it has been the predominant problem in recent philosophy of science. It can hardly be claimed that I have offered a solution to the problem. But I have at least indicated the direction in which the solution seems to me to lie. The guiding principle in the determination of this solution is the fundamental proposition that one cannot employ a scientific entity to explain a perceived entity until he has shown how the scientific entity itself can be defined in terms of perceived entities. The relevance of this general thesis to the point here being considered is that the character of the dictionary relation determines both how the scientific entity explains the perceived entity and how the scientific enity can be defined in terms of the perceived entity.

The *construct* is the totality of those postulates which assert about the content of the suppositional symbol to the extent to which this is inferrible from the character of the descriptive symbols and the nature of the operation. This will be an

[1] Russell, *Our Knowledge of the External World*, p. 101.

extended or a narrow group to the degree to which the operational route is determinate or indeterminate. Co-implication is the foundation for the most highly determinate of these routes, but it is comparatively rare in its occurrence and does not enable us to extend the given in the direction either of the obscurely abstract or of the obscurely concrete. Implication as the passage to greater concretions is highly determinate as to content though it loses something of the existential claim. Implication as the passage to greater abstractions is not so highly determinate as to content but involves no loss in existential claim. All of this has been considered. It was further mentioned that the determinateness of analysis and synthesis is to be estimated by means of a logical theory of whole-part relationships, and that the determinateness of causal inference is to be ascertained by means of a logical theory of temporal complexes. For example, causation is probably not so highly determinate as it has been supposed to be; certainly many of the old philosophical principles such as " cause is like effect," " cause must touch effect," " cause transmits something to effect," do not seem to be borne out by modern science. Considerations of an analagous kind should be introduced with reference to the cases involving other dictionary relations.

The *hypothesis* is the totality of those postulates which assert about the content of the suppositional symbol after it has been determined by the constructional method and increased by the hypothetical method. The methods for the increase beyond the construct are also of two kinds, viz. those which determine its content in such a way as to make the body of postulates implicative of the known descriptive propositions, and those which further determine its content in such a way as to make the postulates implicative of the known propositions and other propositions still to be verified.

We may now see both the structure and the nature of the derivation of a deductive system. Let us symbolize a body of theorems or descriptive propositions by T; then our task is to find a body of postulates, P, about obscure occurrents which are of such a character that they imply T. In order to find such a set we suppose a proposition D which asserts a relation of a certain kind, typical of the realm of clearly given occurrents, between an element of T, i.e. some clearly given occurrent, and an obscurely given occurrent. Then considering T and D as joint premises we infer a body of propositions C; this constitutes that part of the postulate set which we call the construct. Then we set up the relation "DCX implies T," where X designates the additional premises which must be introduced in order to make the implicative relation a valid one. Then we endeavor to fill in the content of X through the hypothetical route, e.g. analogy, imagination, inductive generalization, etc. When such has been done we may call the resulting postulates hypotheses, and they may be symbolized by H. Then DCH implies T. But by further use of the hypothetical method we increase the content of the hypothesis by the addition of H_1, and we then find that $DCHH_1$ implies TT_1, where T_1 represents additional propositions supposedly capable of empirical verification and thus descriptive in character. If they are found to be true through direct inspection of the data we consider that our hypothesis is also true, but if we find them not to be true we reject H_1 and replace it by another group H_2. Then the total group implies TT_2, which is empirically examined in the same way. This indicates clearly what there is of permanence in any hypothesis and why it is not necessary to build again from the ground when the implications of an hypothesis are not verified.

Theories as Probable

It will be well to summarize this discussion of the structure of an explanatory system by raising a specific question. This question is often asked of science, but not always successfully answered. What is meant by saying that a scientific theory or hypothesis is probable rather than certain?[1] Clearly the answer depends upon what one means by "theory." In the broadest sense a theory will be the totality of the postulates, i.e. $DCHH_1$. But one can better see what is meant by characterizing a theory as probable if he attempts to consider in what sense each of the main elements of a theory is probable. Let us attempt this briefly.

[1] Feigl, *Theorie und Erfahrung in der Physik*, pp. 117–129.

The D propositions are all probable rather than certain because they refer to obscure structure. The only guarantee of these structures is the pervasive uniformity of nature. For example if our suppositional symbol is an abstraction from a clearly given occurrent, our proposition D will be true only if the occurrent in question has an abstract associate. We try to minimize this difficulty by letting D represent the most characteristic structures, e.g. associations, abstract-concrete relations, spatial and temporal inclusions, temporal relations, etc.

The C propositions have no greater probability than the D propositions, since they are derived through them. But they have a much decreased probability due to the possibility of erroneous inference through an inadequate understanding of the principles of constructional inference. This is especially true in the case of inferences from parts to wholes and the reverse, and inferences from causes to effects and the reverse.

The H propositions have no greater probability than the C propositions, since they are derived through them. But they have an additional and very significant decrease in probability due to the impossibility of formulating in any precise way the manner in which they are derived. Analogy, imagination, induction and the like do not have any formal guarantees. In fact, often the H propositions seem to take on content with no recognized derivational route at all, as in the cases of sudden insights and flashes of intuition. Hence they lose an important increment of probability.[1] It might be supposed that the H group is also doubtful in the sense that there might be an alternative group, different in content, which would explain the descriptive propositions equally well; hence the H group would not be unique. I am inclined to discount this factor. If the H group does explain the theorems and explains only the theorems (as it is presumed to do at this point), then the data behave in every respect *as if* there were an entity corresponding to the H group; but if they behave *in every respect* as if there were such a group and in no respects contrary to the supposition of such

[1] This leads Russell to assert as the supreme maxim in scientific philosophizing, "Whenever possible, logical constructions are to be substituted for inferred entities," *Mysticism and Logic*, p. 155.

a group, what could be the meaning of asserting the existence of an alternative and different group? The groups would necessarily be *in every respect* alike, and would consequently be identical.[1]

The H_1, H_2 . . . propositions have no greater probability than the H propositions since, again, they are derived through the hypothetical route. But here there is an additional feature of improbability, because we do not yet know the truth of the implicates. Consequently here we not only have an improbability, but we have a method for proving possible falsity, and we cannot therefore assert a very high probability anterior to the application of the test.

It seems likely that what one usually means by the term, theory, is the total group consisting of $DCHH_1$. This has, in general, a low degree of probability. But its structure indicates clearly those elements which are most definitely responsible for the low probability; they are particularly the HH_1 postulates. Hence when a theory proves inadequate we should modify first the H_1 element; if this does not suffice to overcome the difficulty we should then modify the H factor, since some of the original data might have been proved erroneous. The C group will be modified only in unusual cases since this represents the permanent aspect of the theory. The D group will be altered last of all, since this defines the very problem itself; it will be abandoned only in case there is a clear demonstration either of the inapplicability of the structural relation to the situation in question or of the non-pervasive character of some supposedly universal relation.[2]

Adequacy of Suppositional Symbols

The discussion of this concluding topic may be brief since most of what must be said has been anticipated. It is clear

[1] Cf. Poincaré, *Foundations of Science*, p. 141.
[2] The significance of Heisenberg's principle of indeterminacy, or rather, the significance of the conclusion as to the existence of spontaneity in nature which is inferred from this principle, lies in its relevance to this problem. If there are breaks in causal laws, then the structural uniformity of causal complexes is not so pervasive as it has been supposed to be. Hence to define any suppositional symbol through an operational route expressive of causal relations becomes less justifiable, for there may be no cause.

that an explanatory science contains many suppositional symbols. Those who would make rational science non-existential insist that it is made up entirely of such symbols. I have tried to refute this position by insisting that every explanatory science contains a descriptive part which is implicitly if not explicitly present. But it is true that the emphasis of the rational science is upon symbols which do not have any clearly ascertainable referents. We usually say that rational science is concerned with abstractions. This is true, but it inaccurately expresses the relation between rational science and any science which contains suppositional symbols. There are studies which contain many suppositional symbols which are not rational; the most obvious of these is history. Furthermore the distinction between clear and obscure occurrents is relative, hence a science may contain a large number of symbols which appear to be suppositional by comparison with the symbols of another science, and yet which are not essentially suppositional by comparison with other symbols within that same science. For example, sociology may be thought to be suppositional because it deals with society which by comparison with the subject-matter of physics is relatively obscure; but a descriptive science of society is much less obscure than a rational science of society which introduces symbols for social forces, social minds, social wills, and similar highly vague notions. Thus we cannot say that every science which deals with suppositional symbols is rational. But we can affirm that any science which deals with suppositional symbols for abstractions rather than concretions, wholes, parts, past or future, takes on rational organization very readily. We shall discuss this in detail in our next chapter when we consider the integrative value of quantitative notions. Here we wish merely to point out the prominent role which suppositional symbols play in explanatory science.

It appears that the admission of suppositional symbols into any science is the signal for the abandonment of all claims either to precision or to certainty. As Russell suggests, in such a science " we never know what we are talking about nor whether what we are saying is true."[1] We can see what there is of validity in

[1] Russell, *Mysticism and Logic*, p. 75.

this statement. It is true that every theory must be probable rather than certain, since we derive it by an indirect rather than a direct route; it is also true that every suppositional symbol must be obscure rather than clear, since it refers to an occurrent which is given as inherently vague rather than precise. But this does not mean that we have committed ourselves to the abandonment of the basic claims to scientific adequacy. More specifically it does not mean that science must regress back into the theological and metaphysical stages of the Comtian description. It means rather that with the recognition of the unavoidable uncertainty and the inevitable obscurity there is also the recognition of the principles through which both the uncertainty and the obscurity may be reduced to a minimum. These principles are the rules for operational derivation as expressive of basic structures. So long as we accept the act of scientific discovery as an ultimate and insoluble mystery, we shall be committed to the position that scientific theories are not derived but just " happen." But then we shall never know why they explain, what there is in them that enables them to explain, how they must be modified in order that they may explain better, what must be done to them if they do not explain, why they are only probable rather than certain, etc. Again so long as we believe that suppositional symbols have their routes of definition through a direct inspection of obscurely given occurrents rather than through an operational derivation from clearly given occurrents we shall be committed to the position that they are unavoidably obscure. But then we shall never be able to talk about them, or engage in cooperative studies of them, or verify them, or improve upon them, or even reject them. Thus the very conditions of scientific intelligibility and progress involve the recognition of the necessity for operational derivation and definition.

It follows that a science which has admitted suppositional symbols does not cease thereby to be a science. In this respect an extreme positivism is wrong. But a science which admits suppositional symbols without a proper recognition of the methods for their derivation and definition is almost certain to be a bad science. In this respect an extreme positivism is right. A critical positivism will reconcile the two positions by an analysis of operational routes. This will reveal the *norms* for thinking in terms of suppositional symbols, and will thus afford a criterion by means of which adequate suppositional symbols may be distinguished from inadequate ones. The preceding pages have attempted, in outline only, precisely this task. Before concluding we must turn to a consideration of certain methods for explaining descriptive systems—methods which illustrate clearly the advantages of employing suppositional symbols in the portrayal of clearly given occurrents.

CHAPTER XIV

QUANTITATIVE METHODS[1]

In opposition to the extreme positivism to which we have had occasion to refer in the course of the preceding discussion there is another position which is equally characteristic of certain tendencies in modern science. For want of a better term it may be called *mathematical realism*. Though it is a point of view which is more characteristic of the common sense interpretation of science than of science itself, still it has some important defenders among scientists.[2] From this point of view the suppositional entities are not considered unreal or as less real than the clearly given occurrents, as they are for the positivist. On the contrary they are considered more real. Molecules in motion are more real than heat; wave movements in the ether are more real than light; the amorphous space-time is more real than space and time. The assumption is that when any explanatory notion has taken on a fairly specific content and has been shown to be adequate as a principle for explaining a certain group of data, there is no longer any occasion to refer to the data, since the explanatory notion will do all that the data will do. The data then become subjective, or mere shadows of a real, objective, and permanent world.

The essential postulate of this position is that when we have explained anything we have explained it away—though probably few of its defenders would accept the statement in so bald a form. But there is a firm insistence upon that fact that those entities of the word which are called numbers and quantities are more ultimately real than those which are called qualities. Measurement is considered to be the essence of the scientific method. It is employed because quantitative concepts give us an insight into the character of the world which could not

---

[1] Portions of this chapter are reprinted, with important changes, from an article by the author entitled " The Logic of Measurement," *The Journal of Philosophy*, Dec. 21, 1933, vol. XXX, 26.
[2] Jeans, *The Mysterious Universe*, chap. V; Weyl, *The Open World* (New Haven, 1932), pp. 26–9, 84, also *Mind and Nature*, pp. 32, 79–80. The position of Eddington could not be called mathematical realism, though chap. XII of his *Nature of the Physical World* defines the position essentially.

possibly be gained by purely qualitative notions. Quantity is, in fact, fundamentally more rational than quality. A quality is given as a stubborn, brute fact, to be felt or experienced, but not to be comprehended in its given character. A quality is not understood until it is measured. If it can be measured directly, the result can be achieved immediately; if it cannot be measured directly, it must be measured indirectly by correlation with a quality which can be measured; if there is no such quality then recourse must be had to a hypothetical or fictional entity, itself quantitative and supposedly in direct variation with the given quality. Laws, furthermore, which express correlations between two qualities, are not properly laws at all until the qualitative correlations have been replaced by functional relationships of a mathematical kind. Quantity is the key which unlocks reality.

Now as a methodological postulate this can hardly be criticized. But as is often the case in science, a methodological postulate is given the status of a metaphysical judgment. The quantitative aspects of the world are soon looked upon as representative of its essential nature. To explain qualities is to explain them away. To understand them is to be convinced that they are mere appearances. To rationalize them is to construct a system in which they do not function at all as explicit elements. To talk about them is to talk about them vicariously. To grasp them is to realize that they cannot be grasped.

The importance of this notion warrants an examination of it. In considering the problem our purpose will be two-fold: (a) to refute the notion that quantity is more real than quality and that to explain is to explain away, and (b) to illustrate concretely by means of quantity the way in which the discovery of a postulate set increases the integrative character of a descriptive system. I have hinted in earlier chapters that quantitative occurrents belong in the category of relatively obscure occurrents, and that in passing from qualitative occurrents to quantitative occurrents science is undergoing the transformation from the descriptive stage to the explanatory stage. Hence we have reason to believe that there is some specific feature of quantitative occurrents which permits them to be employed so readily as instruments of integration and explanation. Accordingly, we may formulate the problem of this chapter as follows : What is the integrative or explanatory value of suppositional symbols for quantities, and what is the metaphysical implication of the explanatory character of such symbols ?

The complete analysis of the concept of quantity is a problem of some complexity. I must forego this analysis and presume on the part of the reader a general understanding of the notion.[1] Furthermore I shall be obliged to limit my problem in such a way as to exclude all technological questions. This will enable me to avoid all of the problems which have to do with the actual manipulatory activities involved in measurement. To exclude such problems is not to deny their reality. It is rather to acknowledge that their importance is too great to permit adequate consideration in so few pages. Bridgman has well emphasized the importance of these questions in his insistence that the very meaning of the term measurement depends upon what we happen to be measuring.[2]

It is well to be as precise in this discussion as the obscurity of the subject-matter permits. Let us define measurement as the discovery or derivation from an occurrent, through a variety of operations not here specified, of a number which is then spoken of as its measured value. Measurement is the operation of which the outcome is a number and of which the starting point may be called for our purposes a qualitative occurrent. Then when the qualitative occurrent has had a number derived from it and attached to it, it becomes by virtue of that fact a quantitative occurrent. Hence the problem of the explanatory value of quantitative notions is not accurately expressed as the explanation of qualities in terms of quantities ; it should rather be formulated as the explanation of qualities in terms of numbers or measured values. Then one can say that qualities which have been explained in terms of numbers have become quantities. Hence one does not really explain qualities

---

[1] Cf. Albert Spaier, *La pensé et la quantité* (Paris, 1927) ; Russell, *Principles of Mathematics*, part III ; Johnson, *Logic*, Vol. II, chap. VII ; Duhem, *La théorie physique*, pp. 157–70 ; Campbell, *Physics, the Elements*, part II ; Cohen, *Reason and Nature*, pp. 88–99.
[2] *Logic of Modern Physics*, p. 16.

by quantities, he explains qualities by numbers and the qualities thus explained are thereby also called quantities.

Numbers, moreover, are occurrents. Furthermore they are occurrents of the same general kind as non-numerical occurrents. Hence we may say that the symbol "2" is a manner of representing the plurality of a pair of shoes in the same general way that the symbol "black" is a manner of representing the color of either member of the pair. This is true not only of numbers as attributes of classes but of numbers as measured values. But the relation of an occurrent which is a number to the occurrent of which it is the measured value is a relation of a certain kind; it is the relation of abstract occurrent to concrete occurrent.[1] Thus the occurrent 2 which is the measured value of the edge of a table is an abstract occurrent of which the edge of the table is a concretion; this is indicated by the fact that in any situation where the edge of the table is found the 2 will also be found, but the 2 may be found in the absence of the edge of the table. Hence when we attach a measured value to a qualitative occurrent we are simply performing an operation based upon the fact of the structural relation of this occurrent to a higher abstraction, and through the operation we become aware of the latter. But since abstract and concrete occurrents occupy the same reality and the same metaphysical levels of reality neither can be more real than the other. Furthermore the explanatory value of numbers must be due to their abstract character, and they must explain in the way that abstractions always explain concretions. Thus we see in anticipation the solutions to our two problems.

The employment of numerical symbols to explain qualitative symbols may thus be looked upon as a special case of the use of suppositional symbols to explain symbols for clearly given occurrents. It is that case in which the suppositional symbols are derived through the constructional operation of abstraction rather than through concretion, association, analysis, or some other route. But there is a very important difference between the abstraction of numbers and the abstraction of non-numerical occurrents. Non-numerical occurrents of a highly abstract

[1] Cf. Lenzen, *op. cit.*, pp. 26–27; Whitehead, *Science and the Modern World*, chap. II.

sort are relatively obscure, but numerical abstractions are not. This is due to the fact that the latter have become the subject-matter of a highly developed science and have taken on fixed characters and relationships. Numbers for the mathematician are not suppositional symbols but descriptive symbols, and his problem is to explain them in terms of higher abstractions, which are then suppositional. The ordinary investigator who is using numbers as instruments of measurement is not really operating with suppositional symbols at all. Hence it cannot be said that he derives numbers from qualitative occurrents and then endeavors to give to them such content as he pleases in order that from them he may deduce his descriptive propositions. For example, he does not attach the number " 2 " to a given length and then define the number " 2 " through this derivation and through its capacity to imply other known propositions. He cannot make numbers do as he pleases, because the mathematician has already determined what numbers must do; consequently all that he can do is to take over these fixed notions, accept the rules for their organization, and endeavor to determine what possible application they may have to his phenomena. What is suppositional for him, therefore, is not the symbol itself but the legitimacy of supposing that it is a correct abstraction from the occurrent which he is endeavoring to explain through it. Consequently, in order to test this legitimacy, he assumes the relation in question and then on the basis of the fixed character of the number system he endeavors to deduce other propositions belonging within the group of descriptive propositions as already established or as projected into the future. This makes the procedure essentially the same as in the case of obscure suppositional symbols; at least there is the same venturing forth to expand the descriptive system by the assumption of relations not clearly given.

Integrative Value of Numbers

We may now turn to a specific consideration of the problem formulated early in the chapter: What is the integrative or explanatory value of numbers?

If numbers are abstractions they will explain in the same

general way as do all abstractions. Now the characteristic of abstractions is that they permit the transfer of all of their implicates to their concretions. Though the relation of concrete to abstract is not highly determinate, since we may not conclude much as to the character of the abstract from the character of the concrete, the reverse movement is highly determinate. Thus although we can learn little about abstractions when they are defined through the constructional route, we can use them effectively after they have been enlarged through the hypothetical route. There is permitted a generous transfer of features of the abstract system to the concrete system. From the fact of the high integration of numbers and from the fact that numbers are abstractions from the qualitative occurrents which they measure, it follows that the qualitative occurrents take on a higher integration.

This increased integration is exhibited in a number of ways. In the first place it is a commonly recognized fact that a quantitative relation between two kinds of occurrent is a statement of a more significant connection than a qualitative relation. This is true both of correlations between occurrents and relations of resemblance and difference. Qualitative correlations are based upon mere presence or absence of the occurrents concerned, and resemblance and difference are based upon the presence or absence of a quality in one occurrent as compared with its presence or absence in another. As soon as numbers are introduced, presence and absence become replaced by degree of presence and degree of absence, i.e., degree of intensive manifestation; and resemblance and difference become replaced by degree of resemblance and degree of difference, i.e. position in a series. In the second place the introduction of numbers into a system involves a transfer of the structure of the numbers to the structure of the qualitative occurrents. Numbers are derivable from one another by operations, and it follows that qualitative occurrents take on this feature. This gives both the correlations and relations of similarity and difference much less of the character of brute facts, for it is now possible to see why they are as they are. The relations of concrete occurrent to abstract occurrent and of co-ordinate concretes to one another become interpretable in terms of the relations of value to variable and of one value to another. For example, the qualitative relations among red, green, and blue, and between any one of these and the abstract notion of color, are, as purely qualitative relations, somewhat in the nature of surds whose reality we must recognize but whose character we cannot penetrate. One may say in a vague sort of way that the shades differ from one another while at the same time resembling one another in that each possesses the same indefinite relation to the abstract notion of color. When quantitative values representing wave lengths are employed to designate the shades, the nature of the difference between the shades is seen to be of the same kind as that between two numbers, and the nature of the relation of any shade to the abstract notion of color is seen to be of the same kind as that between any number, as a value, and number in general as a variable. In the third place, and this after all constitutes the essential advantage of a number system over a qualitative one, the relational possibilities of the qualitative system are enormously increased by virtue of the extension of the number system to include zero values, negative values, and infinite values, to mention only a few. It then becomes possible to recognize qualitative relations not previously known to exist. For example a circle and a polygon may be shown to resemble one another by virtue of the fact that the circle may be derived from the polygon through the increase to infinity of the number of the sides. So in the same way a straight line may be shown to be a curved line of zero curvature, parallel lines may be shown to be lines which meet at infinity,[1] etc. Thus in general the introduction of quantitative methods permits the substitution of relations of quantitative dependence for relations of qualitative independence or of loose qualitative dependence, and of specific quantitative dependence for relations of general qualitative dependence. Accordingly, a high degree of systematization and integration replaces a low degree, and the aims of rationality are to that extent achieved.

This does not mean, however, that number in general is

[1] This is possible according to Vaihinger (*op. cit.*, pp. 56–61) only through the method of " unjustified transference."

more basic than quality; it asserts rather that the numerical properties of an occurrent are more abstract than the occurrent itself as a quality. In any classification of the categories quantity would probably be given a less fundamental place than quality. Quantity, as a category, is less abstract than quality because it is a species of quality whose differentiating property may be pointed out. Whenever the species of a quality may be arranged into a series by virtue of an asymmetrical, transitive, connected relation which is capable of being interpreted as " greater than," this quality may also be spoken of as a quantity. Thus there are quantitative and non-quantitative qualitites, and a quality becomes a quantity when certain conditions are fulfilled. Furthermore in a logical system, such as that of the *Principia Mathematica*, quantitative notions are introduced late in the derivation; qualitative notions are presumed to be logically more basic, since one can derive quantitative notions from them. Hence in accepting the explanatory value of quantitative notions one is not committed to the position that qualities are superficial rather than basic features of the world.

Implications of Quantitative Explanation

Our second problem is that of showing the error involved in drawing certain consequences from the explanatory property of numbers. The adequacy of the system of numerical postulates in explaining and increasing the integration of the body of descriptive propositions leads one to attribute to number a representational or substitutional character which is apt to be quite misleading. One feels that since the descriptive propositions are contained by implication in the postulates there is no longer any occasion to refer to them specifically. And from this it becomes but a short step to the conclusion that since they are there only implicitly they are really not there at all. Hence the world becomes simply and purely mathematical, and we may completely describe it in terms of numbers and functional relations.

There are many objections to this position, and I shall briefly discuss several. In the first place from a set of postulates which is *merely numerical* in character one cannot deduce the body of descriptive propositions. One can make this deduction from the complete set of postulates but the numerical propositions are only a part of this set. They are that part which states the interrelations of the numbers among themselves; i.e. they are that part which we called in the preceding chapter, the hypothesis. But the hypothesis is useless without the dictionary, for from the hypothesis alone we can infer only further features of the hypothesis. With the addition of the dictionary, however, we can get from the realm of suppositional symbols over into the realm of descriptive symbols. Thus from the equation for falling bodies,

$$s = \tfrac{1}{2} g t^2 + v_0 t$$

one cannot infer anything until he knows that it is the equation for falling bodies and that the various symbols s, g, t, v_0 refer respectively to space, gravitational attraction, time, and original velocity. But this is given only through the dictionary. Hence in explaining away quality by reducing it to quantity one reduces it rather to quantity and to the dictionary, but since the quality must be in the dictionary one has not really explained it away. In other words, he has admitted its existence in trying to explain it away.

In the second place there is the obvious fact that every quantity is a quantity of something. There are no mere numbers in a system of quantitative postulates but lengths, areas, volumes, durations, velocities, masses, etc. The representation of quantities by numbers tends to obscure this fact, for in the numerical representation we often lose sight of the unit which is necessarily involved, and whose qualitative character makes our measurement truly representative. Thus when we attach a number to a length we do not dispose of the length once for all by replacing it by a number; we substitute for it a numbered length, and the problem remains. The same situation prevails, though in a different manner, in those cases where we do not measure directly. To measure the quantity of heat by the length of a column of mercury is not to reduce quality to quantity, but merely to substitute for a quality to which numbers cannot be readily attached another quality to which they can. The "reduction" of heat to molecular motion is, in principle, the same, for motion is still a quality, i.e. something qualitatively distinct from, say, angular spread.

In the third place number is itself a quality. Numbers are so readily arrangeable into series possessing quantitative features that we tend to forget their unmistakable qualitative character apart from such arrangement. Number is a given kind of thing just as is color or sound. Certainly number cannot be distinguished from color in any quantitative way; yet it is distinguishable, and it must be so in a qualitative way. So also two and three are qualitatively distinguishable though we find it hard to hold our attention on the qualitative distinction since the quantitative one seems so much more obvious. They seem to be qualitatively distinct species of the genus number, distinguishable from one another in the same general manner that red and green, or bitter and sour are distinguishable.

The objection which one is tempted to make at this point is somewhat as follows : But do we not institute measurement precisely in order that the mysterious relationship between red and green may be made intelligible by substituting for each of them a number representing its wave length, and thus showing that the relation between red and green is the same as that between two numbers, and that the relation between red and green, respectively, and the general notion of color is the same as that between two specific values of a variable and the variable itself ?

Such an objection begs the very question at issue. Since the relation of two numbers to one another is simply the relation of one of two co-ordinate species to the other, and since the relation of any value of a variable to that variable is merely an instance of the species-genus relationship, it is hard to see why talking in terms of numbers helps us to understand colors. We simply have two qualitative systems which are structurally similar; if a relation is mysterious in one system, it will be equally so in the other. Apart entirely from ordering relationships, it seems that the qualitative difference between twoness and threeness is just as ultimate and inexplicable as the qualitative relation between redness and greenness, and although we seem to be able to *derive* twoness and threeness from number in general as a matrix, we are actually able to do this with no more success than we are able to derive redness and greenness from color.

In the fourth place we must not overlook the fact that, however great may be the structural similarity of the two systems, as we pass from the descriptive propositions to the postulates we lose the essential *quale* of the quality to be measured. Hence we cannot successfully get back from the quantitative system to the qualitative system. Where the measurement is of a very direct sort this need not be true, provided we do not lose sight of the unit. But in the less direct sorts of measurement, where we correlate colors with wave lengths, and tones with vibrations, and heat with columns of mercury, it is clearly absurd to suppose that the measured value in any sense replaces the quality measured. When qualities are measured they are not explained away. There is thus no reason to suppose that the reduction of heat to quantitative terms makes it unreal or subjective or in any way removes it from the realm of empirically given entities. Quality, apparently, is closer to sentiency, just as quantity is closer to rationality. But there seems to be no reason why sentiency should be absorbed in rationality.

The fifth objection is the general incapacity of a quantitative system to represent dichotomous divisions, i.e., to handle two-valued systems. Where qualities manifest themselves not by degrees but by complete presence or complete absence, there can be no quantitative representation. Thus it is impossible to show quantitatively how two qualities may be at the same time similar because species of the same genus and yet contradictory because implying contradictory differentia. Bogoslovsky's attempt[1] to explain qualitative differences in terms of a principle of quantitative indices applies to divisions which are only *apparently* dichotomous and is utterly futile when forced upon genuine dichotomous divisions.

In the sixth place one must insist that measurement is in many cases an affair of so artificial a character, that one is presuming a great deal to suppose that a knowledge of the number system conveys any essential information with regard to the qualitative system. Many qualities may be ordered as intensive quantities

[1] B. B. Bogoslovsky, *The Technique of Controversy* (London, 1928), p. 139.

and then more or less arbitrarily correlated with the system of ordinal numbers. As a result the system of qualities may be supposed to take on the relationships of the ordinal numbers. One of the best examples of this sort of thing is the measurement of hardness upon the basis of the asymmetric, transitive, connected relation " scratches." When the serial arrangement has been completed, one member of the series is chosen in a completely arbitrary manner as the standard of hardness, and the correlation of the ordinal numbers begins at this point. It is easily seen that the numbers represent the qualities not inherently but only in the sense that they locate the qualities in a graded series of similar qualities. Thus the informational value of such a representation is dependent upon the choice of a standard value, the compactness of the series, the regularity of the intervals between adjacent members, and the terminating members of the series. And in any case the only structural relations which may be transferred from the system of quantities to the system of qualities are the ordinal relations. These are definitely more limited in scope than the cardinal relations.

One might easily prolong the argument,[1] but I stop with the mere mention of certain further limitations, such as our inability to solve mathematical problems beyond a certain degree of complexity so that the mathematization of qualities may involve only a pseudo-rationality. Again there are all of the difficulties associated with measurement as it is actually carried on, what Ritchie[2] calls the inherent error of measurement, the fact of the fluctuation of values so that we are compelled to be satisfied with averages, means, modes, etc., the fact that all measurement rests ultimately upon a judgment of sense either in the reading of the recording instrument or in the construction of the recording instrument in the first place, etc.

Such considerations as these contribute to the refutation of the claim that the quantitative aspects of the universe are any more basically real than the qualitative. Numbers and quantities seem to permit integration in a way in which qualities do not. This is a fact, and an important one. For by means of the structural

[1] Cf. d'Abro, *Evolution of Scientific Thought*, pp. 411–26 ; Aliotta, *Idealistic Reaction Against Science*, part II, chap. IV ; Ritchie, *Scientific Method*, chap. V.
[2] *Op. cit.*, p. 128.

relations of qualities to quantities we may explain the former by the latter. In this way qualities become more intelligible by exhibiting their relations to other entities and by sharing the structural relations of these other entities. But the qualities remain. We have no less in the world than we had to start with. Qualities are neither explained away nor reduced. They remain as part of the ultimate furniture of the world.

But the alternative claim is equally mistaken. We cannot suppose that the world is merely the sum total of obvious qualities. Many occurrents are given obscurely. This is a fact, and an important one. For by means of the structural relations of the obvious occurrents to the obscure ones we may explain the former by the latter. In this way the obvious occurrents become more intelligible by exhibiting their relations to other entities and sharing the properties of these other entities. But the obscure occurrents are not added to the scene by a fictive act of mind. We have no more in the world than we had to start with. Constructs and hypotheses are features of the world which are discerned through operational analysis. They also are part of the ultimate furniture of the world.

It would indeed be surprising, would it not, if science could make the world other than it is?

INDEX

An entry listed in the singular number ordinarily refers to the plural as well, and vice versa.

Where divisions are made under an entry, the main divisions are set off with semicolons; subdivisions of these are separated by commas only; occasionally further divisions are marked by parentheses. Items in a division are usually arranged alphabetically according to a principal word.

Abstract, 74, 77, 101; reality of the, 319; structure of abstract-concrete relations, 197 n.; symbols as, 217, 241; *abstraction*, 80, in explanatory science, 301, extensive, 127–8, 218, 230, quantitative, 278, in rational science, 313, from series, 127; *abstractions*, 28, 59, 76, 77, 81, 196, as defined in terms of concretions, 217–220, 304, as explanatory of concretions, 319–320, as unconscious discoveries, 128; *abstractive method*, 211, 215, 217–220, operation, 217–218, route, 218; *abstractness*, 24–25, 74; see *Occurrents, abstract*.
Acquaintance, 107 n., 108
Activities, 17–26; classification of, 18–26; conflicts between, 34; human, 17–26, 33; science as a value activity (*see Science*); value activities, 17–26, 50–55
Alexander, S., 48 n.
Aliotta, A., 249 and n., 327 n.
Analogy, closeness of, 229; *analogical operation*, 228–230
Analysis, 136, 220, 276; qualitative, 73 n.; spatial and temporal, 78; validity of, 78, 136, 166; *analytic method*, 105, operation, 222, propositions (*see Propositions*); *analyzed occurrent*, 160
Anti-positivists, 5, 283
a posteriori, 22, 248, 296

a priori, 22, 248, 292, 296, 300
Aristotle, 41 n.
Assertion, 161, 180, 183–184
Association, 66–71, 73, 74, 173; general features of, 69–71; repeated, 186, 191–192, 215–222; *associative method*, 212, 215–216; *co-associates*, 70, 197; see *Occurrents, associated*; *Complexes, associative*
Attention, 98, 273–275, 295–297
Avenarius, 5
Awareness, 46, 47, 48, 49–50, Ch. V; of applicability, 110–116, 233; complex forms of, 96; content of, 47, 49, 94, as derived, 118–119, as developing, 117, independent, 94, kind of, 47, knowable, 95, non-symbolic, 48, 49, objective, 95, rational, 95, secondary, 123, 130, symbolic, 48–50; control of, 99–105, 115; dependence on content, 102; distinctness and vagueness in, 98; grounds for, 46, 53; intensity of, 98; kinds of, 47, 99–102, 105–116; locus of, 102–105, 117; of meaning, 99–100; of non-symbolic occurrents, 100–102, 114; not transforming, 94–96; objects of, 97; of occurrents, 99, 102, 233, 270; of particularity, 109; of reference, 158; of referent, 99; relation to content, 94–98; as

329

selective, 102–105, 272–275, 295–297; successive, 118; of symbolic occurrents, 102, 111, 112, 114; of transition, 117, 119, 121; undifferentiated, 105; *awareness situation*, 105
Axioms, 303, 304; see *Postulates*
Ayres, C. E., 17 n.

Baconians, 293
Barry, F., 58 n.
Bavink, B., 5, 62 n., 136 n., 211 n., 223 n., 264 n., 283 n.
Benjamin, A. C., 223 n., 316 n.
Bergson, H., 5, 110
Berkeley, 243
Bliss, H. E., 17 n., 27 n.
Bogoslovsky, B. B., 326
Boole-Schroeder logic, 307
Boyle's law, 214
Bradley, F. H., 105 n.
Bridgman, P. W., 5, 139 n., 145 n., 229 n., 249 n., 318
Broad, C. D., 5, 36 n., 39 n., 62 n., 218
Buchner, 5
Burtt, E. A., 17 n.

Campbell, Norman, 58 n., 159 n., 186 n., 207 n., 275 n., 305, 318 n.
Cantor, G., 218
Carmichael, R. D., 197 n., 209 n., 223 n., 241 n., 242 n., 288 n., 292 n., 302 n.
Carnap, R., 5, 58 n., 60 n., 77 n., 78 n., 81 n., 85 n., 87 n., 136 n., 149 n., 219 n.
Cassirer, E., 211 n., 218 n., 249 n., 273 n.
Categories, 23, 28, 36–37, 159, 323; in science, 40
Cause, 165, 309; *causal laws*, 128–129, 277, 312 n.; see *Laws ; Statistical generalization*
Chamberlin, T. C., 210 n.
Charles' law, 214

Class, 80–81, 153–155, 188, 195, 221, 243; extensional, 81, 155, 221; in intension, 155; *classification*, in explanatory science, 301, of sciences (see under *Science*); *classificatory stage in science*, see *Science*
Clausius, R., 214
Cognition, empirical analysis of, 54–55; aspects of, 47; elements in, 55; scientific, 45–60; *cognitive situation*, 48, 49, 113; see *Knowledge*
Cohen, M. R., 39 n., 223 n., 249 n., 264 n., 282 n., 298 n., 300 n., 302 n., 318 n.
Co-implication, 186–187, 188–191, 196, 215–216, 228, 297
Compatibility, 177–186, 196, 277
Complexes, Ch. IV; determinateness of, 164–167; structural features of, 172; associative, 73–77, 80–81; 163, 165, 174–175, 221, (see *Association*); dissociative, 77–81, 163, 165–166, (see *Dissociation*); correlational, 163, 174, 196, 297, types of, 166, (see *Correlation*); operational, 117; *relation-complex*, objectivity of, 137, reflected by operation, 135, (see *Operation*); *complexity*, 24–25, 66, degrees of, 65, kinds of, 65; see *Occurrents, complex*
Comte, A., 5, 24, 238, 242, 314
Concepts, 159–162, 179–186; compatibility of, 183; complex, 184; conjunctive union of, 180; content of, diversity in, 182, modifiability in, 180; existence claim of, 161–162, 180; extensional reference of, 182; as principle of individuation, 183; intension of, 169, 180, 182; of metaphysics, 75; negative of, 162; of science, 197, 219; structural relations between, 180; symbolic relations

between, 162; *conceptual formulations*, 249

Concrete, 75, 77, 81, 101; reality of concrete and abstract, 319; relation to abstract, 191-192, 225; symbols as, 241, 300; *concreteness*, 74; concretion, structural relation to abstraction, 302; *concretions*, 28, 76-77, 197, as defined in terms of abstractions, 220-221, 305; as unconscious discoveries, 127-128; *concretive method*, 212, 215, 216-217, 220-222; see *Occurrents, concrete*

Conjunction, 228; see *Compatibility*

Constancy, 61; *hypothesis of*, 59

Constructs, 46, 77, 80, 89-90, 126-128, 198, 199, 209, 211-212, 214, 218, 283, 328; definition of, 218; *"the" construct*, 307, 308-309; construction, 5, 198, 288, logical, 219, 283, 311 n., symbolic, 195; constructional definition, 213-214, method, 133-134, 211-222, 223, 231, operation, 271, route, 59, 212-222, 308, symbols, 199, 213, 224, 231, (in mathematics) 213;

Contain, 73-74, 78

Content, see under *Awareness; Occurrent; Symbol; Symbol, meaning of; General characterizing symbols; Concept; Proposition*

Contradiction, 175-176, 193-197; definition of, 194; operation of, 195

Convention, 283, 305

Correlation, determinateness of, 164; intimacy of, 164; qualitative, 278; of qualities, 316; quantitative, 297; repeated, 165, 297; statistical, 277, 297; *correlational structure*, 222; see *Occurrents, correlated; Complexes, correlational; Correlational symbols*

Correlational symbols, 162-168, Ch. VIII, 199, 231, 235-236, 285;

adequacy of, 276-283, 297-302; derivation of, 277; in descriptive science, 277-283; determinateness of, 191, 196; in explanatory science, 297-302; kinds of, 164, 236, (see *Incompatibility, Contradiction, Nonimplication, Compatibility, Implication, Co-implication*); operational origin of, 172; universal, 189, 277; general laws, derivation of, 170 n.; *generalizaton*, 195, *justification of*, 170 n., 231; *Generalized correlation, justification of*, 170 n.

Cramer, F., 133 n.

d'Abro, A., 283 n., 327 n.

Dampier-Whetham, W. C. D., 213 n., 214 n.

Darwin, C., 133

Data, see *Given; Science, data of*

Dedekind, R., 218

Deduction, 227; *deductive organization of science*, 242, structure, 298, 302, systems, 247, 302-310, (relation to descriptive) 303-310

Definition, 44, 220-221, 303; of abstractions, 217, 304-305; of concretions in terms of abstractions, 220-221, 304-305; of constructs, 218; constructional, 213-214; through denotative gesture, 280; in descriptive science, 280-281; in explanatory science, 298-300; extensional, 280, 285; principles justifying, 304-305; method of, 220-221; *of obscurely given*, 83, 88, 90, 116, 295-296; operational, 219, 289, 314; structural, 281; of suppositional symbols, 216, 217, 219, 288; redefining, 299; see *Symbols, derivation of*

Denotative gesture, 43, 58, definition through, 280; *denotative method*, 91, 208, 217; *denotative reference*, 91

Description, 171, Ch. XII; as analysis, 276; as explanatory, 286; in explanatory science, 290–297, 313; knowledge by, 107 n.; as naming, 286; *descriptive stage in science*, (see *Science*); see *Descriptive knowledge*; *General characterizing symbols*
Descriptive knowledge, descriptive science, 33, 241–242, 246–247, Ch. XII; adequacy of, Ch. XII; change in, 279–280; correlational symbols in, 277–283; definition in, 280–281; dependence of content on occurrent, 287; descriptive symbols in, 278–281, 286; expansion of, 287–289; experiment in, 270–272, 282; explanation in, 282; exploratory direction in, 281–282; extension in, 246; gaps in, 281–282, 285; instability of, 268; logical priority in, 274–275; obscurely given occurrents in, 284–286; unchecked observation in, 279–280; organization of, 278–281; connection with practical, 246; prediction in, 280; relation between symbols in, 279–280; principle of selection in, 281; structural relations in, 278, 281; suppositional symbols in, 283–289; see *Description*; *General characterizing symbols*
Dictionary, the, 305–308, 324; dictionary relations, 307–308; translation, 305
Difference, 65; qualitative, 66, 68, 81; quantitative, 66; structural, 66; see *Otherness*; *Diversity*
Dingle, H., 58 n., 211 n.
Discovery, 89, 103, 119–134, 296; abstraction as, 128; conscious, 124–130; discovery or invention of hypotheses, 303–304; of occurrents, 269–275; obscurely given, 200–202; operations of, 270, 272, 274; discovery or invention of postulates, 303–304; scientific, 132–133, 223, 250, 288; symbolic, 49; symbols as, 126; unconscious, 121, 128, 130–134
Dissociation, 66–71, 71, 73, 74, 175; extensive and minute, 78; general, features of, 69–71; repeated, 186; see *Occurrents, associated*; *Complexes, associative*
Diversity, 70, 77, 182; qualitative, 70; spatio-temporal, 70; symbolic, 176–177, 196; see *Otherness*; *Difference*
Duhem, P., 249 and n., 250 n., 252, 263, 318 n.

Eaton, R. M., 73 n., 141 n., 149 n., 242 n.
Eddington, A. S., 5, 84 n., 145 n., 208 n., 249 n., 259 n., 316 n.
Elements, 73, 77, 78, 160; kinds of, 81; of propositions, 184; spatial or temporal, 78; in symbolic system, (see *Symbolic system*)
Empirical, 25; laws, 304, 305; origin of meaning, 138; theory of meaning, 57, 175; science, 241, 268, 272, 277–278, 284; character of symbols, 46, 57–58; derivation of symbols, 108
Empiricists, English, 97
Ends, 17–18; see *Values*
Enriques, F., 187 n., 249 n., 283 n.
Error, 97, 114
Esthetics, 37
Ethics, 37–38
Evaluation, 25
Event, 62; see *Occurrent*
Everett, W. G., 20 n.
Existence, 56, 219; constructional, 284; factual, 84 n., 284; of meaning, 105, 137, 148; of occurrent, 196, 296; in rational science, 296; of referent, 148;

INDEX 333

existent, 125; *existential claim*, (see under *Symbol, Concept, Proposition, Suppositional symbols, Hypothesis, Operation*); *existential science*, 240, 296

Experience, science as regulative of, 299–300; *value experiences*, 17–26, (see *Activities*)

Experiment, 56; in descriptive science, 270–272, 282; in explanatory science, 293; imaginative, 257; *thought-experiment*, 231

Explanation, 46, 205, 207–211, 212, 222–232, 286–287, Ch. XIII; and description, 268, 286–287; in descriptive science, 282; as explaining away, 316, 317, 323–328; in explanatory science, Ch. XIII; principles justifying, 304; kinds of, 208–210; method of, 46; pseudo-explanations, 210, 215; quantitative, Ch. XIV; of science, 57–60; verbal, 210, 215; *explanatory notions*, 283, value of numbers, 319–323, value in science, 284, (in suppositional symbols) 318; see *Explanatory knowledge*

Explanatory knowledge, explanatory science, 242 and n., 247–248, 271, Ch. XIII; abstraction in, 301; adequacy of, Ch. XIII; classification in, 301; correlational symbols in, 297–302; definition in, 298–300; relation to descriptive, 303–310; descriptive part in, 313; descriptive symbols in, 292-297; dependence of elements in, 298; experiment in, 285; explanation in, 302; as directive of exploration, 301; as filling gaps, 300–302; growth in, 298–300; hypotheses in, 301, 305–307, 309–311; instability in, 298; integration in, 247; intension in, 247; obscurely given occurrents in, 291, 294, 297, 301–302; operations in, 293–294; structural relations in, 298; suppositional symbols in, 306–310, 312–315; systematic revision in, 298–300; transition to, 278, 281, 289, 292–293, 298, 303; see *Explanation; Rational science*

Extension (logical), 154–155; in descriptive knowledge, 246; *extensional definition*, 280, 285; see *Symbol; Symbolic system; Concept; Class*

Extension (spatial or temporal), 78, 80, 81; *extensive*, 78–81, 101, 129; minute, 78–81, 101, 129

Facts, 84 n., 142, 199, 235, 269.
Faraday, M., 249
Feigl, H., 187 n., 224 n., 249 n., 281 n., 310 n
Fictions, 126–127, 198, 209, 211 n., 284; operations in deriving, 126–127; summational, 275 n.
Flint, R., 17 n., 36 n.
Form, of symbol, of meaning, see under *Symbol*

Gay-Lussac, 213
General characterizing symbols (characterizing symbols, descriptive symbols), 137, 148, 151–162, 176, 195, 206, 235, 243, 244–245, 267, 285; adequacy of, 268–276, 292–297; content of, 151, 286; in descriptive science, 278–281, 287; in explanatory science, 292–297; instances, 151, 153; intension of, 173; invention in forming, 275–276; kinds of, (see *Concepts, Propositions*); possibility as a character of, 158; potentiality expressed in, 152; in rational science, 292–296; reference, determination of, 255, indirectness of, 258; referents of, 151,

152, 156; *generality in symbols*, 152, 153–157, 158, 168, 176, 235, 244; *generalization*, 151–154, 176, 195, 275–276; characterization, 152–153, 155, 157, 267, positive, 157, negative, 157; *generalized characterization or description, descriptive characterization, description*, 46, 151, 171, Ch. XII; see *Description*; *Descriptive knowledge*

Generality, of science, 24, 28; in symbols, 152, 153–157, 158, 168, 176, 235, 244.

Generalization, through general characterizing symbols, 151–154, 176, 195, 275–276; through correlational symbols, 195, 231, justification of, 187 n., 231, operation of, 231; *statistical generalizations*, 304; see *General characterizing symbols*; *Correlational symbols*; *Hypotheses*; *Laws*

Genus, 182, 228

Geometry, 292, 304; non-Euclidean, 298; theorems of, 304

Given, 6, 7, 71, 82, 83, 96, 219; *clearly given*, 6, 80, 83–86, 89, 92–93, 131, 239, 243, 247, 272, 275, determinateness of, 103, independence of, 103, relativity of, 281, 286, 313; *obscurely given*, 7, 57, 83–86, 92–93, 128–130, 240, 328, definition of, 83–84, 90, exploration of, 288, reduction of, 85, 87, relations connecting with clearly given, 86–87, 90; *givenness*, 108; see *Awareness*; *Occurrents*

Goblot, E., 17 n.

Gore, G., 223 n.

Grounds, of applicability of symbol, 115; for awareness, 46, 53

Haeckel, E., 5
Heat, kinetic theory of, 214
Heisenberg, 312 n.

History, compared with science and philosophy, 27–31; *historical object*, 81

Hobson, E. W., 5, 202 n., 211 n., 249 and n., 273 n., 275 n., 283, 298 n.

Homogeneous, 80–81, 188; homogeneous-heterogeneous, 80–81

Hypotheses, 46, 77, 80, 179, 199, 205, 209, 214, 221, 288, 328; *ad hoc*, 227; discovery or invention of, 303–310; existential claim of, 225; in explanatory science, 301; conversion into fact, 199; limitation on, 288; modification in, 312; in Newton, 211 n.; probability of, 229, 310–312; specificity of, 232; in Vaihinger, 211 n.; "the" hypothesis, 305–307, 309–310; *hypothetical entities*, 198, 283, heuristic value of, 283; *hypothetical method*, 134, 211–214, 222–232, 309, route, 212–214, 309; *hypothetical symbol*, 148, 197, 224, existential claim of, 225

Icons, 148, 149, 153, 230, 242, 243, 250; determinateness of, 259; as portraying individuality, 254–255; intension in, 173; meaning of, 149; reference of, 149, directness of reference, 254–255, 258; in relation to simplification, 259; specificity of, 230; verifiability in, 255; *iconic character in word symbols*, 254–255; *iconic symbolism*, generality in, 252–254, 258; *iconic system*, 244; see *Images*; *Models*; *Pictorial knowledge*

Idea, 96–98; metaphysical status of, 97–98

Ideal, 248, 292, 296; *idealizations*, 127, 198, 229–230, 283

Ideals, 17–21, 20 n., 32, 34

Identity, 66, 68, 77

INDEX

Images, imagery, 149, 250–252; concreteness of, 259; flexibility in, 256–258; generic, 243; as leading to working hypotheses, 257; incompatibility in, 263; inadequacy for structural organization, 265; lack of syntax in, 262; *image systems*, instability of, 261–263, incapacity to portray obscure occurrents, 259–261; *imageless thought*, (see *Thought*); see *Icons*; *Models*; *Pictorial knowledge*

Imagination, 101; creative, 122; *imaginative experiment*, 257, route, 230–231

Implication, 186–191, 196, 217–222, 297–298, 302, 309

Inclusion, 63, 70, 74, 277

Incompatibility, 175–176, 186, 187, 188, 192, 196, 297; symbols for incompatibilites, 264–265

Indeterminacy, principle of, 312 n.

Indices, 149–151; as extensional only, 173

Individual, 25; individuality, 255, of occurrent, 63, 152, 154, 275; *principle of individuation*, 71, 183

Inductive inference, 226; syllogisms, 231; *inductive and deductive science*, 240

Inference, 164, 166, 174, 209, 212, 215, 226, 228; analogical, 229 n.; causal, 128, 221; empirical foundation of, 184; inductive, 226; probable, 206

Infinity, 218, 219, 322

Ingression, 156 n.; see *Participation*

Instrumental techniques, 269–270, 270–271; *instrument, principle underlying*, 270

Intension, see under *Symbols*; *General characterizing symbols*; *Concept*; *Class*; *Propositon*; *Explanatory knowledge*; *Symbolic system*

Intuition, 108–110, 111; in Bergson, 110

Invention, 89, 102–103, 119–133, 135, 137, 296; in forming characterizing symbols, 275–276; conscious, 121, 123, 124–127, 128, 134; of hypotheses, 303–304; of postulates, 303–304; results of, 135; of suppositional symbols, 200–202; of symbols, 104, 137, 168, rules for, 138; symbols as, 125; unconscious, 121–123; *inventive operation in symbols*, 275–276, 297

Isolation, 273–275

James, W., 251 n.

Jeans, Sir James, 5, 71 n., 84 n., 264 n., 316 n.

Johnson, W. E., 62 n., 318 n.

Joseph, H. W. B., 73 n.

Joule, J. P., 214

Judgment, 112, 114; metaphysical, 317; of sense, 327

Kelvin, Lord, 249, 264

Keynes, J. M., 181 n.

Keyser, C. J., 240 n., 242 n., 302 n.

Knowledge, 44, 94; adequate, 56, 94, 241; *a posteriori* and *a priori*, 240; control of, 115–116: development of, Ch. X; descriptive and explanatory, 240; empirical and rational, 240; justification of, 54; object of, 48, 55; origin of, 238; real and ideal, 240; representative and non-representative, 241; symbolic character of, 115; task of, 238; of the world, 72; knowing, 56; see *Cognition*; *Descriptive knowledge*; *Explanatory knowledge*

Krutch, J. W., 17 n.

Langford, C. H., 302 n.

Laws, 71, 84 n., 186–187, 277, 297, 317; causal, 128, 277, 312 n.; derivation of, 187 n.; empirical,

304, 305; empirical foundation for, 187; probability of, 187 n.; *statistical generalization*, 277, 297, 304
Lenzen, V., 145 n., 218, 273, 305 n., 319 n.
Levy, H., 273 n.
Lewis, C. I., 88 n., 145 n., 300, 302 n., 307
Locke, J., 308
Lodge, O., 249
Logic, 166; Boole-Schroeder, 307; non-Aristotelian, 298; of science, 38, 40, Ch. II, method of, 45–46; principles of, 301; symbolic, 143; *logical form*, 255

Mach, E., 5, 89 n., 231 n., 249
Martineau, H., 238 n.
Matter, 218, 220
Maxwell. C., 249
Meaning, see *Symbols, meaning of*
Measurement, 271, 316; artificiality of, 326; definition of, 318; inherent error in, 327; outcome of, 318; starting-point of, 318; *measured value*, 318, 324, 326; see *Number ; Quantity*
Memory, 100, 102.
Metaphysics, 36–37; concepts of, 75; of science, 40, 219; *metaphysical division*, 73 n., judgment, 317
Method, 22–23, 25, 40, 56, 216, Ch. XI, XII, XIII, XIV; abstractive, 211, 215, 217–220; analytic, 105; of anticipation, 295–296; associative, 212, 215–217; case, 277; concretive, 212, 215, 217, 220–222; constructional, 133, 211–222, 223, 231; of definition, 220–221; denotative, 91, 208, 217; of generalization, 231; hypothetical, 134, 211–232, 309; of interpretation, 295–296; of a logic of science, 45–46; quantitative, Ch. XIV; suppositional,

293; *methodological assumptions*, 304, postulates, 317; see *Operation*; *Experiment*
Meyerson, E., 5, 84 n., 211 n., 283 n.
Mill, J. S., 5, 187 n., 242 n.
Minute, 78–80, 101, 129
Models, 243–244, Ch. XI; material, 250; see *Pictorial knowledge*
Montmasson, J. M., 223 n.
Moore, G. E., 5, 48 n., 105 n.
Morris, C. W., 145 n.

Names, 149–151; application of, 150; proper, 150, 173, as securing directness of reference, 275; naming, 171, 173, 286–287
Nature, 56, Ch. III–IV, 115–116; specific properties of, 61; structure of, 71, 163; structural features of, 206; uniformity of, 88, 206; *natural*, 82; see *Occurrents, realm of*
Negation, 158–159; negatives, negative *symbols*, 176, 194–195; *negative numbers*, 322; see *Symbols, negative reference ; General characterizing symbols, negative characterization ; Concepts, negative of ; Propositons, negative of*
Neptune, 210
Newton, Isaac, 211 n., 219
Nicod, J., 187 n.
Non-existential, 248, 292, 296, 313
Non-implication, 177–180, 185, 196, 277, 278
Non-pictorial knowledge, 244–245; simplification in, 244–245; see *Descriptive knowledge ; Explanatory knowledge*
Norm, 33; *normative science*, 33
Number, numbers, 192, 219, 318–319; abstract character of, 319; as abstractions, 319–320; as attributes of classes, 319; constructional derivation of, 319; as descriptive symbols, 319;

explanatory value of, 319–323; integration of, 321, integrative value of, 320–323; as measured values, 319, 323–325, 326; mere, 324; negative, 322; as occurrents, 319; ordinal, 327; structure of, 321; *number system, extension of,* 322; *numerical, qualities,* 323, 324–325; *numerical symbols,* as *suppositional,* 320; *infinity,* 218, 219, 322; *relation of value to variable,* 322; *zero,* 322; see *Quantity ; Measurement*

Object, 44, 57, 62, 73, 75; of awareness, 97, (see *Awareness*); fixity of, 58; historical, 81; independence of, 58; of knowledge, 48, 55; rationality of, 57, 58–60; scientific, 85 n.; sense, 85 n.; of symbol, 48, 55, 57, (see *Symbol, referent of*); unreal, 97; see *Objectivity ; Occurrent*

Objectivity, 61; in awareness, 95; hypothesis of, 59; of operation, (see *Operation*); see *Object*

Observation, 56, 101, 105–116; in descriptive science, 279–280; symbolic expression for, 115; *observer,* 90

Occupant, 193; *occupy,* 67

Occurrent, 47, 48, 56, Ch. III–IV, 94, 143; absence of, 67; abstract, 73–74, 75, 181, 319, (see *Abstract*); analyzed, 160; associated, 163, 180–181, 184, (see *Association*); awareness of, 99, 102, of character of, 107, production in awareness, 233, 269; clearly given occurrents, 89, 116, 235, 237, 239, 267, obvious, 274–275; cognitive, 47; complex, 65, 66, Ch. IV, 160–161, 184, (see *Complexes*); concrete, 73–74, 181–183, 217, 319, (see *Concrete*); content of, 63, 112, 114, 144, 146,

152, 154, 168, 170, 234, 288, structural, 63, 148, unique, 63, 65, 66, 68, 148, 171; correlated, 163; discovery of, 269–275; dissociated, 184, (see *Dissociation*); existence of, 296, form of, 63, 112, 144, 167; general, 153–154; inclusion of, 63, 70, 74; individuality of, 63, 152, 154, 275; kinds of, 71, 99, 155, 177, 235, 244, 267, 275, determination of, 106; locating of, 150; location of, 66, 71, 75; metaphysical status of, 101; non-relational, 73; non-symbolic, 99–102, 115, 139, 144, 148, 154; obscurely given occurrents, 89, 103, 116, 207, 233, 235, 237, 239, 247, 284, 294, content of, 91, 92, 146, 166, 196, determination of content of, 166, definition of, 88, 90–91, 296, derivation of, 89, in descriptive science, 284–286, as discovered, 200–201, existential status of, 196, in explanatory science, 291, 294, 297, 301, 302, inherent obscurity in, 233, 294, location of, 89, meaning of, 88, operation through which obtained, 88, 89, reduction of, 92, relation to clearly given, 88, 89, 206, route through symbol, 137; overlapping, 63, 70; particular, 243, particularity of, 152, 154, 235; properties of, 65, 71; qualitative, 318, 321, (see *Quality*); realm of occurrents, 65, 67, 70, 83, 196, plurality in, 176, structure of, 65, 73, 82, 162, 196; relational, 64–66, 73, (see *Relation*); simple, 74, 76, 160, 184; spatio-temporal, 66, spatial, 67, 71, 74, temporal, 67, 74, (see *Space and time, Space, Time*); structure of, 160, 163, 164, 173, 184, 189, 190, 205,

235, 236, 247, 288, (see *Structure*), spatio-temporal, 66, 160, structural relations of, 66, 72, 171-172, 175, 197, 206, 237; symbolic, 99–100, 102, 111, 112, 115, 139, 140, 144, 145, 148, 154; unanalyzed, 160; *occurrential system*, 114; *sub-occurrent*, 47, 73, 75; see *Given*

Ogden, C. K., 140, 281 n.

Operation, 88 n., Ch. VI; abstractive, 217–218, 271, as negational, 217–218; analogical, 228–230; analytic, 222; awareness of, 123; conscious, 118–119; constructional, 271; of contradiction, 195; definition of, 118; of derivation, 57, 120–121, 137, 139; of discovery, 270, 272, 274; of duplicative construction, 244; existential claim of, 296; exploratory, 274; of generalization, 231; kinds of, 118-121; dependence of meaning on, 144–145, 148; normative aspect of, 134–138; objectivity of, 121; from occurrent to symbol, 138; of particularization, 150; performance of, correct, 135–136, rules for, 134–138; physical or manipulative, 269–272, 284, 318, in explanatory science, 293–294, (see *Instrumental techniques*); public, 89, 90, 101, 120, 124, 125, 127, 128, 129, 135, 137, 296; and relation, 130, 135, as equivalent to relation, 91; repeatable, 88–92, 93, 101, 120, 124, 125, 127, 128, 129, 135, 137, 296; result of operation, 89, 121, 123, content of, 135, 136; of selection, 272–275, 285, 295–297; spatial, 221; structure of, 117–118; symbol of, 137; from symbol to symbol, 137; synthetic, 222; transforming, 273; unconscious, 118–119; validity of, 137; *operational complex*, 117–118, definition, 289, 314, derivation, of symbols, (see *Symbols*), of suppositional symbols, (see *Suppositional symbols*), entity, 89, 90, origin, foundation, of symbols, (see under *Symbols*); *operational route*, 56, 91, 92, 101, 125, 136, 145, 147, 168, 171, 179, 192, 194, 202, 203, 205–207, 237, 315, abstractive, 218, determining applicability, 233, constructional, 59, 211–222, 308, determinateness of, 190, 205, 289, hypothetical, 211–214, 309, imaginative, 230–231, expressing structural relation, 205, through symbol, 136; *operational techniques*, 61, 133, habitual, 131; *operational theory of meaning*, 57, 139; see *Selection; Analysis; Synthesis; Definition; Inference; Contradiction; Invention; Discovery*

Ostwald, W., 249

Otherness, 65, 66, 68, 176–177; qualitative, 176; spatio-temporal, 176; see *Diversity; Difference*

Parker, D. H., 70 n.

Participation, of symbol in situation, 156–159, 167, 174–175, 177–178, 183–185, 188, 193; joint, 178, 183, 188; repeated, 186–187; manner of reference of, 167; referent of, 167; *non-participation*, 156, 158–159, 175–176, 177–178, 185, 188, 193; *participation and non-participation, repeated*, 186–187

Particular, 62, 63, 101, 142, 154; *particularity*, 150, 235, awareness of, 109, of occurrent, 152, 154, 235, 236; *particularization*, 150; *particular proposition*, (see *Proposition*)

INDEX

Pearson, K., 5, 89 n., 264 n., 283
Peirce, C. S., 17 n., 101 n., 149 and n., 250 n.
Perception, see *Observation*
Perfect entities, 229–230
Phenomenalism, 83
Philosophy, 36–39, 77; compared with science and history, 27–31; critical, 36; of life, 21; of science, 36–41; speculative, 39
Physics, mathematical, 305; theoretical, 305
Pictorial knowledge, 242–244, Ch. XI; pictorial and non-pictorial, 241, 251; *pictorial symbolism*, advantages, 256–258, 266, disadvantages, 258–266, substitutional character, 256; *adequacy of pictorial system*, 254–255; see *Models*; *Images*, *imagery*; *Icons*
Planck, M., 5, 85 n., 283 n.
Poincaré, H., 5, 84 n., 132, 224 n., 249, 273 n., 283, 305 n., 312 n.
Positivism, 238–239, 314, 316; critical positivism, 6, 314; *positivists*, 5, 89, *Vienna group*, 5; *positivistic attitude*, 270, 287, 314, 316; *positivistic stage in science*, 238–239
Possibility, 158
Postulates, 303–305; discovery or invention of, 303–310; methodological, 317; numerical, 323; proper, 304–310; *postulate set*, 317; *postulate system, relation to descriptive*, 304–310
Pragmatists, 5
Prediction, 226–227; in descriptive knowledge, 279–280
Principia Mathematica, 323
Principles, of logic, 304; of natural sciences, 304; of procedure, 304.
Probability, 187 n.; of hypotheses, 229, 310–312; of laws, 187 n; of theories, 310–315, 314; *probable inference*, 206

Proper name, 150, 173; as securing directness of reference, 275
Proposition, 159–162, 180–181, 183–185; analytic, 75–76, 180; as asserting, 180, 183–185, facts, 160; compound, 184; content of, structural, 183–185, unique, 183–185; in descriptive science, 279; elements of, 184; existence claim of, 161–162; in explanatory science, 297; general, 189; intension of, 170, as representing intension of concept, 183; negative of, 162; particular, 277, 304; relations among, 185; symbolic relations between, 162; simple, 185; singular, 277, 304; synthetic, 90–91, 180; system of, 184; universal, 189, 223; *propositional interpretation*, 307
Psychology, 35; behavioristic, 35
Publicness, 58, 59, 61; hypothesis of, 59

Quality, 71, 316–317, 322, 326; explanation of, 318–319, 327–328; intensive, 326; as relational, 276, 287; species of, 323; *qualitative analysis*, 73 n., diversity, 70, similarity, 66, 68, 80; see *Occurrents, qualitative*
Quantity, 272, 316–319; as species of quality, 323; and quality, reality of, 327–328; *quantitative abstraction*, 278, aspects of world, 317, correlations, 297, explanation, Ch. XIV, (metaphysical implications of) 317, 323–328, methods, Ch. XIV, occurrent, 318, relations, 321, (as degree of presence) 321, representation, 326, symbols, 285; see *Number*; *Measurement*.
Quantum mechanics, 298; phenomena, 71 n.

Rankine, W. J. M., 211.
Rational, 25, 57, 58; rationality, construct of, 59, hypothesis of, 59, of object, 57, 58-60, in science, 242, 284, 313; rationalistic theories of meaning, 57; see Rational science.
Rational science, 241, 268, Ch. XIII; abstraction in, 313; dependence on descriptive, 290-291; descriptive symbols in, 292-297; existence in, 125; integration in, 297-300; see Explanatory knowledge.
Realists, 5; new, 5; critical, 5; mathematical realism, 316.
Reality, 27, 43, 62, 64, 85 n., 95; real, 82, 90, 101-102, 106, 296.
Realm, 82-83, 86, 89; determination of, 106; sub-realm, 86; see Occurrents, realm of; Symbols, realm of
Reference, referent, see under Symbols
Reflection, 21, 33, 117, Ch. II; critical, Ch. II; study of, 36, 38
Relation, 63-66, 178, 196; abstract-concrete, 191-192, 197 n., 225, 302; as determining attributes, 160; as connecting, 160; content of, 143; external, 64; functional, 317, 323; internal, 64; obscurely given, 87, 90; of obscurely given to clearly given, 86-87, 90; operation and, 91, 130, 135; quantitative, 321; spatial, 72; temporal, 72; structural relationship, 136, between concepts, 179, of concretion to abstraction, 302, in descriptive science, 278, 281, in explanatory science, 298, and operational route, 205, of symbols, (see under symbols); relational possibilities, 322; see Occurrent, relational; Structure; Operation

Relativity, theories of, 280, 298
Religion, 34-35, 37
Ribot, Th., 223 n.
Richards, I. A., 140, 281 n.
Richardson, E. C., 17 n.
Rignano, E., 223 n., 231
Ritchie, A. D., 58 n., 82 n., 159 n., 186 n., 211 n., 327
Ross, W. D., 41 n.
Rueff, J., 227 n., 242 n., 292 n.
Rumford, Count, 214
Russell, B. A. W., 5, 17 n., 62 n., 70 n., 85 n., 89 n., 107 n., 142, 209 n., 218, 221, 242 n., 259 n., 283, 308 n., 311 n., 313, 318 n.

Schlick, M., 5
Science, 43-44, 49; adequacy of, 53, 54, 270; aim of, 61; aspects of, 49; categories in, 40; classificatory stage in, 269; compared with philosophy and history, 27-31; composite, 35; concepts of, 197, 219, operational definition of, 219; constructs of, (see Constructs), without constructs and hypotheses, 198; conventions in, 283, 305; as creative, 138; data, 77, 80, 82-86, 85 n., 316, collection of, 269, selection of, 83-84; deductive organization of, 242; empirical, 241, 268, 272, 277-278, 284; existential and non-existential, 240, 296; explanatory value in, 284; exploratory stage in, 269; fictions in, 126-127, 198, 209, 211 n., 284, operations in deriving, 126-127, summational, 275 n.; generality of, 24, 28; goal of, 52; hypotheses in, (see Hypotheses); idealizations in, 127, 198, 229-230, 283; inadequacy of, 53, 54; inductive and deductive, 240; laws in, 71, 84 n., 128, 186-187, 277, 297, 304, 305,

INDEX

312 n., 317, (see *Laws*); logic of, 38, 40, Ch. II, method of, 42–46; broader meaning of, 52–55; metaphysics of, 40, 286; method in, (see *Method*); methodological assumptions of, 400–401; nature of, 15; normative, 33; phenomenological analysis of, 47–50; philosophy of, 36–41, 219; affiliation with philosophy, 36; place of, Ch. I, historical or functional, 17; positivistic stage in, 238–239; principles of, 304; general problem of, 138; rational character of, 313, rationality in, 284; as regulative of experience, 299–300; science (of sciences), 26, 31, 40, 42; stages in, 242, Ch. XI–XIV; starting point, 71, 77, 80, 83; statistical correlation and generalization, 277, 297, 304, (see *Statistical generalization*); structure of, Ch. II, rational structure in, 242; subject-matter, 23–24, 32, 42, 61, 62, 83, 199, dependence of, 24; suppositional symbols in, (see *Suppositional symbols*), in descriptive science, 283–289, in explanatory science, 306–310, 312–315; task of, 35, 61, 82, 87–93, 116, 233; theoretical entities in, 198, 209, 283; theories in, (see *Theories*); value in, positive, 51–52, 53, negative, 51, 52; as a value activity, 50–55; *Sciences*, classification of, 22–36, (principles of) 24, 25, interrelations of, 39, table of, 32–36; *scientific discovery*, 132–133, 223, 250, 288, (see *Discovery*); *scientific entities*, 308, defined in terms of perceived, 308, as explantory of perceived, 308; *scientific objects*, 84 n., symbolism,

249–250; see *Pictorial knowledge; Descriptive knowledge : Explanatory knowledge; Rational science*
Scientific method, see *Method*
Selection, in descriptive knowledge, 281; habitual, 295; operation of, 272–275, 285, 295–297; unconscious, 295; *awareness as selective*, 102–105, 272–275, 295–297
Sense-datum, 62; *sensa*, 218; *sense objects*, 84 n.
Sense perception, see *Observation*
Series, 127, 323; abstraction from, 80; limits of, 127, 229–230; position in, 321; *serial extension*, 229–230, existential claim in, 230; *serial organization*, in explanatory science, 301
Similarity, degree of, 80, 228; qualitative, 66, 68, 80; spatio-temporal, 80, 228
Situation, 67–68, 70, 72, 74, 226, 227, (see *Space and time*); of occurrent, 67; participation of symbol in, 156–159, 167, 174–175, 177–178, 183–185, 188, 193
Space, 62, 66, 67, 71–72, 73, 77, 220; remote in, 129; *spatial operations*, 222, relations, 72; see *Space and time; Occurrents spatial*
Space and time, omnipresence of, 71–72, 73; privileged position of, 71–72; as structural complexes, 72; structural properties of, 73; *spatio-temporal diversity*, 71, similarity, 81, 228; *spatial or temporal analysis*, 78, elements, 78, extension, 78, 80, 81, extensive, 78–80, 101, 129, minute, 78–80, inclusion, 63, 70, 74, 277; see *Space; Time; Occurrents, spatio-temporal, spatial, temporal*
Spaier, A., 318 n.
Spearman, C., 223 n.

Species, 182, 228; relation to genus, 192
Spencer, H., 5
Stace, W. T., 5, 60 n., 85 n., 218, 264 n., 284, 286 n.
Statisitical generalization, 304; *statistical correlation*, 277, 297
Stout, G. F., 251 n.
Structure, 82; correlational, 222; deductive, 298, 302; kinds of, 236; of nature, 71, 163, 205–206, (see *Occurrents*); obscurely given, 236; repeated, 187, 297; of science, (see *Science*); of structure, 184, 278; *structural adequacy*, (see *Symbolic system*); *structural content*, (see *Content*)
Studies, 24–36; humanistic, 25–26, 32, 34, 35, 37
Subject-matter, of science, see *Science*
Subsistence, 154, 219
Substance, 62
Suppositional symbols, 148, 179, Ch. IX, 235, 236, 245, 316; for abstractions, 313; adequacy of, 283–289, 312–315, criterion for, 315; advantages of, 315; relation to clear occurrents, 209; definition of, 200–207, 288–289; relation to descriptive, 206; in descriptive systems, 283–289; existential claim of, 206, 215, 225, 227; in explanatory science, 306–310, 312–315; explanatory value of, 318; function of, 207–211; in history, 313; as invented, 200–202; in mathematics, 212; numerical symbols as, 319–320; origin of, 200–207, 212, 231–232; operational derivation of, 200, 202–203, 205–206, 212, 216–222; for quantities, 317; *suppositional method*, 293; see *Constructs ; Hypotheses*
Symbolic system, system of symbols, 114, 145–146, 169–170, 176–179, 233,

247, 298; adequacy, 54, 233–238, 240, 241, 267–268, difficulties in ascertaining, 233-234, criteria of, 234–238, 267–268, degrees of, 233–234, extensional, 237, intensional, 237, isomorphic, 237, kinds of, 245, stages of, 234, structural, 236; consistency within, 289; elements in, place of, 298, symbols as, 145, underived and derived, 302–305; expansion of, 288–289; function of, 252; goal of, 239; instability of inadequate system, 234; integrative character in, 317, degrees of, 322; locus of, 239; modifiability of, 54; plan of, 239; representative character of, 235; as simplifying, 252–254; structure of, 207; as substitute, 252; degree of systematization in, 322; principle of division for systems, 240–242; see *Deductive system ; Symbols*
Symbols (symbolism), 46, 55–56, 83, 90, 92, 97, 101, 108, 139, 145; as abstract, 217, 241; adequacy of, 138, (see *Symbolic system*); applicability of, 54, 57, 114, 233–234, awareness of, 110–116, 233, grounds for, 114, operational route determining, 233, symbol for, 48, 55–56, 94, tests for, 113; application of, 108; applicational, 275; a priori character in, 300; classified as to predominant kind of content, 203–205; for clear occurrents, 92, 137, 268; as concrete, 241, 300; content (of symbols), 114, 139, 142, 144, 145, 146, 148, 152, 155, 167, 170, 234, derivation of, 171, determination of, 137, 138, 286, development of, 170, difference in, 143, modifiability in, 170, structural, 146, 148, 170, 171,

176, 203–205, 237, 285, unique, 146, 148, 169, 171, 177, 203–205, 237; constructional, 199, 213, 224, 231; correlation in, repeated, 188; derivation of, 57, 137, 171, 205, 233, empirical, 57, 108, from occurrents, 138, 141, operational, 126, 146, 168, 314, rules for, 314; derived, 49; (see *Definition*); as discovery, 126, symbolic discovery, 49; diversity in, 176–177, 196; empirical character of, 57; existential claim of, 147–148, 159, variability in, 147–148; extension of, 145, 148, 153, 276; form of, 142, 144, 146, 148, 151, 167–168, 171, 234; for future, 100; generality in, 152, 153–157, 158, 168, 176, 235, 244; hypothetical, 148, 197, 224, existential claim of, 225; for incompatibilities, 264–265; indefiniteness in, 233; intension, 145, 148, 153, 155, 163, 171, 179, 185, 186, 266, derivation of, 171, operational origin of, 172; invention of, 104, 137, 168, as invention, 125, inventive operation in, 275-276, 297; for justification, 55, 56, 94; general kinds of, 143; logic of, 143; meaning, 48, 49, 87, 91, 109, Ch. VII, VIII, IX, 234, analysis of, 139–144, awareness of, 99–100, 111, 141, content of, 112, 142–144, 148, 153, (structural, 63, unique, 63), dependence on referent and operation, 145, 148, derivation of, 56, 57, determination of, 105, 137, 148, empirical origin of, 138, existence of, 148, form of, 112, 142–143, 148, 155, nature of, 145, of non-symbolic, 100, of obscurely given occurrent, 88, pragmatic definition of, 145 n.,

meaning property, 141, 143–148, psychological aspect of, 141, as reference, 143, as way of referring, 140, as a relation, 143, meaning situation, 140, meaningful symbol, 159, 179, theories of, (empirical, 57, 175, of Ogden and Richards, 140, operational, 57, 139, rationalistic, 57), ways of, 143, 154; multiplicity in, 243–245; negative (negatives), 176, 194–195; numerical, 319–320; object of, 48, 55, 57; for obscure occurrents, 91, 92, 116, 136–137, 179, 189, 205, 245, connection with clearly given, 66, content of, 146, definition of, 91, 92, 116, general form of, 146, meaning of, 163; occurrential foundation for, 140, 141; for occurrents, 94; for occurrential structure, 173; for operation, 137; operational foundation of, 168; participation (in situation) of, (see *Participation*); for past, 100; pictorial, 256–266; as possibility, 158; quantitative, 285; realm of, 184, 198, plurality in, 176; reference (of symbols), 48, 92, 143, 148, 167, denotative, 91, direct, 146, 148, 179, 195, 235, 237, 244, (means of securing) 275–276, generality of, 235, 275, indirect, 146, 148, 179, 195, 235, 237, 245, 275, negative, 157–159, 195, (awareness of, 158, existential claim of, 159), to occurrents, 234, positive, 157, 195, principle of symbolic, 125, spatial and temporal, 72, ways of referring, 142, 156, (general, 146, 168, 235, 255, specific, 146, 168, 235, 236, 255); referent, 97, 101, 125, 139, 141, 142, awareness of, 100, content of, 142, existence of, 148, kinds of, 142, locating of,

255; referential character of, 104, 108, 140; for relation, 92, 174; relational, 162; relations between, symbolic relations, 92, 163, 169, 173–174, 195, in descriptive science, 280–281; structural relations of, 143, 148, 171, 172, 178, 179, structure of, 162–164, symbolic structure, 177; similarity in, 228; symbolic situation, 140, 141; as social entities, 126; as substitute, 235; types of, 148, 167; typical differences in, 143; symbolic verification, 49, (see *Verification; Truth*); see *Occurrents, symbolic; Symbolic system; Constructs, constructional symbols; Hypotheses, hypothetical symbols; Suppositional symbols; Indices; Icons; General characterizing symbols; Concepts; Propositions; Correlational symbols*

Synthesis, 79, 136; validity of, 166; synthetic operations, 222; synthetic propositions, (see *Proposition*)

System, 81; deductive, 247, 302–310, relation to descriptive, 303–310; occurrential, 114, (see *Occurrents*); postulate, 304–310; of propositions, 184; see *Structure; Symbolic system*

Terms, 64; as designating elements in facts, 160

Theorems, 303; of geometry, 304

Theoretical entities, 198, 209, 283

Theories, 288; probability of, 310–312, 314

Thing, 73, 75, 246; see *Object*

Thinking, 47, 141; scientific, 138; vagueness in, 104; in words, 249, 251; see *Thought*

Thompson, J. A., 186 n.

Thought, 99–102, 110–112, 117; image and imageless, 241, 251; thought-experiment, 231; see *Thinking; Reflection*

Time, 62, 66, 67, 71–72, 77, 286; temporal relations, 72, succession, 277; *past and future*, knowledge of, 100, as invented or discovered, 128–129; *past*, 90; see *Space and time*

Translation, 305

Truth, 20, 34, 112–113, 114, 233, 303; marks of, 54; tests of, 54, 56, 113; see *Verification; Symbols, applicability, adequacy; Symbolic system, adequacy*

Tyndall, J., 133, 264 n.

Universal, 101, 142, 153–154, 234; propositions, (see *Proposition*); see *General characterizing symbols*

Universe, 74, 79

Universe of discourse, 194

Vaihinger, H., 124 n., 198 n., 211 n., 275 n., 322 n.

Value, 20; activities, 17–26, 43; experiences, 17–26; situation, 50; *evaluation*, 25; *science as a value activity, values in science*, (see under *Science*)

Variable, relation of value to, 322

Verification, 49, 112, 113, 114, 179, 205, 254, 310; positive, 159; negative, 159

Wallas, G., 223 n.

Weiss, P., 302 n.

Weyl, H., 224 n., 316 n.

Whitehead, A. N., 5, 17 n., 39 n., 62 n., 78, 78 n., 85 n., 101 n., 127, 156 n., 187 n., 199 n., 218, 283 n., 319 n.

Whole, 75; organic, 81; reduction of, 78; and part, 78–80

Wittgenstein, L., 5, 62 n., 63 n., 109 n., 142 n., 151 n., 160 n.

World, external, 219; of physics, relation to world of sense, 308

Young, J. W., 302 n.

Zero, 322

For Product Safety Concerns and Information please contact our EU
representative GPSR@taylorandfrancis.com
Taylor & Francis Verlag GmbH, Kaufingerstraße 24, 80331 München, Germany

www.ingramcontent.com/pod-product-compliance
Lightning Source LLC
Chambersburg PA
CBHW071759300426
44116CB00009B/1142